Greener Solvents in Conservation

An Introductory Guide

Greener Solvents in Conservation

An Introductory Guide

Edited by Gwendoline R. Fife

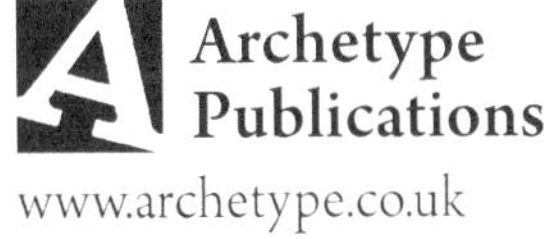

www.archetype.co.uk

in association with Sustainability in Conservation

First published 2021 by Archetype Publications Ltd in association with
Sustainability in Conservation

Archetype Publications Ltd
c/o International Academic Projects
1 Birdcage Walk
London SW1H 9JJ
www.archetype.co.uk

ISBN: 978-1-909492-84-4

British Library Cataloguing in Publication Data
A catalogue record for this book is available from the British Library.

Sustainability in Conservation is extremely grateful to Tru-Vue Inc. for its very generous support of this publication.

Designed and typeset by Marcus Nichols at PDQ Digital Media Solutions Ltd, Bungay, Suffolk

Contents

Foreword

It gives me great pleasure to introduce this handbook to launch the 'Sustainability in Conservation Greener Solvents' project. While fully acknowledging that the conservation practitioner faces an increasing variety of specifically developed, as well as other, materials, this book focuses on the more traditional materials of choice – organic solvents. For certain treatments, where their use cannot be avoided, organic solvents remain indispensable, and appropriate solvent selection requires consideration not only of how they will interact with the planned conservation materials, but also how they might potentially affect the materials in the original substrate and their impact on human health and the environment. Since any one of these aspects can be a complex issue in itself, the incorporation of greener solvent approaches in practice can appear a challenging and even overwhelming objective.

The aim of this handbook is therefore to start removing the hurdles step-by-step. After a fundamental assessment by the conservator as to whether solvents can be avoided in the first place, the next most impactful action is eradicating the current and widespread use of the most harmful solvents. Our primary need to identify and eliminate the use of these potentially most harmful solvents is therefore addressed by providing supporting information on the most straightforward ways to determine the alternatives (ideally already available in the studio) that may be implemented in their place. Given that for many conservators, time and resources are stretched, this can also provide a convincing approach for those who understandably wish to limit the introduction of frequent adaptations in their practice. It is unrealistic to aim for zero impact on our surroundings, but we can seek to minimize it with greener solutions, and the ongoing trends in solvent legislation should encourage us towards this positive and realistic perspective.

I hope that this publication will serve to provide students, practitioners and researchers with introductory information and insights into further research needs and that it will leave them with an irrepressible urge to start taking 'greener steps' in their solvent use today.

Gwendoline R. Fife
Director, Greener Solvents Project

Note: A PDF of this book is available free of charge at www.siconserve.org/greener-solvents/greener-solvents-hand-book/ and on the Sustainability in Conservation website.

Acknowledgements

The level of interest in this resource and project, and the exceptional support offered by so many over the last months has been enormously appreciated and demonstrates the timely importance of information and further research on this topic. I would like to thank Caitlin Southwick for her initial vision in creating the 'Sustainability in Conservation Greener Solvents' project to support and disseminate research into the use of greener solvents in conservation, and her unwavering support and positivity. I am also very grateful to the previous SiC project team members including Aline Assumpção, Thea Christophersen, Bianca Gonçalves, Mariana Escamilla Martinez and Lucile Pourret, for proposing a handbook, and current team members Aline Assumpção, Anabelle Camp, Marie Desrochers, Izabella Gill-Brown, Karoline Sofie Hennum and Lucile Pourret are especially thanked for their energy and hard work in ensuring its creation. I would like to thank all the contributors to this publication for their consistent application and generous input. The excellent advice and suggestions provided by the Scientific Review Committee have been indispensable, as has the assistance of the editorial team, including sub-editors Annabelle Camp and Izabelle Gill-Brown, and authors for their implementation. Heartfelt thanks are extended to the team at Tru Vue for providing sponsorship and the team at Archetype Publications for their assistance.

Although these acknowledgements relate directly to this handbook, we are of course incredibly indebted to many others, including those who have already pledged their continuing personal and institutional participation in this project.

Scientific Review Committee

Prof. Tom Welton, Professor of Sustainable Chemistry, Imperial College, London, UK

Shayne Rivers, Subject Leader, MA Collections Care and Conservation Management, West Dean College of Arts and Conservation, Chichester, UK

Dr Bronwyn Ormsby, Principal Conservation Scientist, Conservation Science and Preventive Conservation, Tate, London, UK

Dr Rosie Grayburn, Scientist at Wintherthur Museum, Garden and Library, Affililiated Associate Professor, University of Delaware, Delaware, USA

Gwendoline R. Fife, GRF Art Conservation Consultancy, Hoegaarden, Belgium

Dr Fergal Byrne, CEO, Addible, York, UK

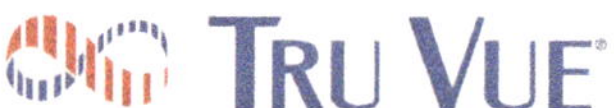

List of contributors

Aline Assumpção, Museu de Arte de São Paulo Assis Chateaubriand, São Paulo, Brazil (alineasd.oliveira@gmail.com)

Fergal Byrne, Addible, York, UK (fergal.byrne@addible.co.uk)

Marie Desrochers, Marie Desrochers, Utah Division of Arts and Museums, Utah, USA (mdesrochers@utah.gov)

Matthew Eckelman, Department of Civil & Environmental Engineering, Northeastern University, Boston, Massachusetts, USA (m.eckelman@northeastern.edu)

Gwendoline R. Fife, GRF Art Conservation Consultancy, Hoegaarden, Belgium (gwen.fife26@gmail.com)

Karoline Sofie Hennum, Independent, Portsmouth, UK (karolinesofiehennum@gmail.com)

Melissa Lewis, Rachel Barker Associates, Bendon Valley, London, UK (melissa@rachelbarkerassociates.com)

Lucile Pourret, Independent, Paris, France (pourret.lucile@gmail.com)

Justine Wuebold, UCLA/Getty Interdepartmental Program in the Conservation of Cultural Heritage, UCLA, Los Angeles, USA (jwuebold@g.ucla.edu)

Sub-editors

Annabelle Camp, Winterthur/University of Delaware Program in Art Conservation, Wilmington, Delaware, USA (annabellefcamp@gmail.com)

Izabella Gill-Brown, National Museums Liverpool, Liverpool, UK (izabella.mckie@gmail.com)

Introduction

Izabella Gill-Brown and Gwendoline R. Fife

Conservators will utilize solvents throughout their careers in a variety of settings and situations. With their use prevalent in general conservation practice, it is therefore necessary to consider how solvents can affect human health, the environment and, not least, the artwork. Their obligatory use alongside concern for our environment has led Sustainability in Conservation (SiC) to prepare this guide and thereby provide those who are seeking greater sustainability in their conservation practice an opportunity to explore the basics of greener solvent usage. Appropriate solvent implementation requires specialist knowledge and is dependent on discipline. While this handbook is intended to be a general resource for all members of the conservation community, paintings conservators are noted, anecdotally, to use more solvents than others in their daily work; accordingly the focus has been placed from this perspective.

Shades of green

Since the term 'green' is often used arbitrarily, the need for a clearer definition is also acknowledged. Just as a solvent can only be described as 'good' or 'strong' with respect to a specified solute, so a solvent's 'greenness' is also comparative to another. To emphasize this relative nature, SiC has opted to employ the term 'greener' rather than 'green' solvents since this highlights the comparativeness of an assessment and hopefully helps prevent the

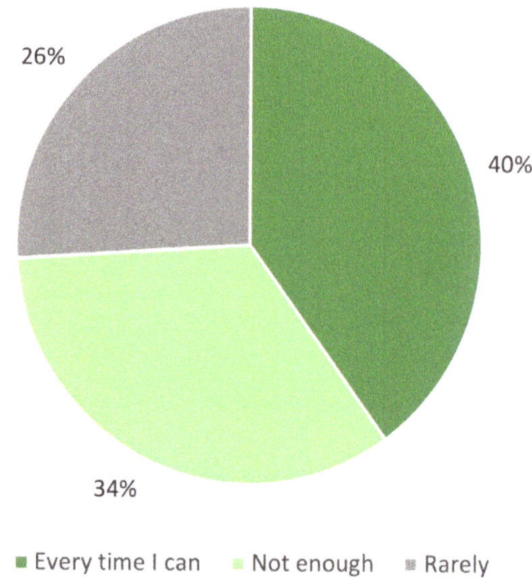

Figure 1 Survey data showing how often conservators use 'green' solvents. Within the survey the term 'green' was left undefined.

spread of misinformation. This change in terminology has been similarly applied by other companies and providers of solvents. 'Greener solvents' can be defined as *solvents that provide a better alternative to other solvents in use in terms of human toxicity, safety, and environmental impact,* therefore the term 'greener' and this associated definition will be used throughout this handbook. Certain traditional solvents and mixtures currently being used may already present a relatively greener solution, while others warrant replacement as soon as possible. How to both assess and apply a greener solvent approach is further clarified within this handbook in Chapters 1 and 3.

The need for such clarity is evidenced by a 2020 international survey (conducted by SiC for internal use only) designed to examine current thoughts on, and usage of, green(er) solvents within the conservation field (Fig. 1). The survey responses to the question of usage reflected the respondents' awareness of what green(er) solvents are and the gap in strong definitions and information regarding them. The majority (60%)

of the 72 respondents explained that they do not use green(er) solvents because of a lack of reliable and accessible information (Fig. 1). Although there are useful guides that provide a framework for defining greener solvents in chemistry and industrial research fields (Capello 2007; Prat *et al.* 2016), specific resources are required which consider the requirements and operating environment within the conservation profession.

The aims of this handbook

This handbook aims to assist those conservators who would like to adopt a greener practice in their work and suggests a basis for further research. Through discussion of some of the history, the critical concepts and consideration of the solvents currently available, we hope to enable greener solvent selection and encourage broader interest and research into this subject.

The principles of green chemistry provide a basis for the emergence and development of modern (neoteric) greener solvents. Highlighting a history of greener solvents clarifies the comparative nature of their identification and provides insights into the potential and safe application of greener solvents (both traditional and modern) within the conservation field for both the practitioner and artwork. Moreover, the combination of background information with practical guidelines will help conservators to identify and eradicate in the first instance their use of the worst solvent offenders based on safety, health and environmental ratings.

Considerations in solvent use: health, safety and the environment

Solvents are used in many ways to produce and maintain global requirements for products relating to everything from industrial and medicinal needs to everyday items used in the home. As conservators, we use solvents for a variety of tasks. They are often integral to our practice, yet solvent exposure can cause direct harm to the human body. Given the imperative health and safety aspects, *the importance of scientific training for the correct handling of solvents and hazardous materials cannot be overstated.* As conservation has a foundation in the physical sciences, there exist established health and safety frameworks applied within the training programmes such as the use of personal protection equipment (PPE) and safety data information. Although conservators try to select the least harmful option for both the artwork and themselves, more information needs to be collected on solvent usage and greener alternatives in order to make clearer greener choices. Since the unneccessary, improper use or disposal of these materials can incur damage to both practitioners and the environment, greater alignment between the training of scientists and conservators could be considered. How we have come to consider and assess these issues has evolved, and a historical perspective on the development of solvent safety and 'greenness' within conservation is examined in Chapter 2.

Reducing and assessing environmental impacts

To combat our impacts, SiC is working in alignment with the UN's 17 Sustainable Development Goals, a set of interlinked goals which form a collective blueprint for a more sustainable future. We can also look to green chemistry, a relatively new field of research stemming from the late 20th century.

Green chemistry was formalized by Anastas and Warner's 1998 book, *Green Chemistry: Theory and Practice*, in which 12 principles of green chemistry were defined (Fig. 2). These principles are broad and applicable to the entire chemical industry, demanding the prevention of waste, design and use of safer chemicals, implementation of energy and material efficient processes, and the valorization of renewable feedstocks. The Royal Society of Chemistry summarizes green chemistry as 'the utilisation of a set of principles that reduces or eliminates the use or generation of hazardous substances in the design, manufacture and application of chemical products'. This area of chemistry is dedicated to developing and experimenting with environmentally friendlier materials, including many of the solvents that are currently being used throughout the world on a large scale.

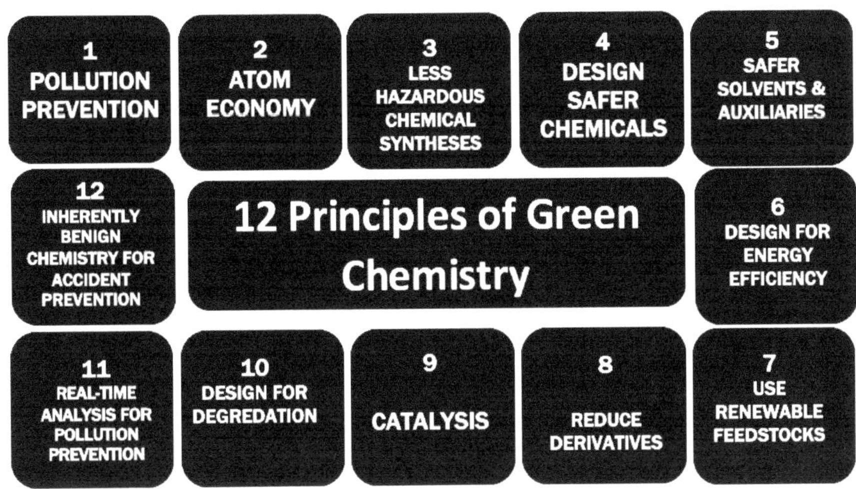

Figure 2 The 12 principles of green chemistry.

Ideally, solvents could be derived from non-food biomass resources and would be benign during use and end-of-life. However, solvent performance is also a key factor. Jessop (2017) states that the ideal green solvent 'is a solvent that makes a product or process have the least environmental impact over its entire life cycle'. As such, the raw materials used (percent bio-based content) and the health, flammability, aquatic toxicity and atmospheric breakdown properties of the solvent must be balanced with the overall performance of the solvent in the desired application.

The performance of the solvent can be assessed in many ways depending upon the application. For example, in synthetic chemistry, a faster rate of reaction might be observed, or the reaction may be carried out at a lower temperature. Alternatively the products can be more easily isolated when one solvent is used over another. In the coatings industry, the viscosity or the solvent-polymer mixture is vital to ensure even coating with no stretching or bubble formation. A conservator may similarly assess solvent performance in their coating applications, alongside the critical consideration of potential effects on the original substrate.

Fergal Byrne

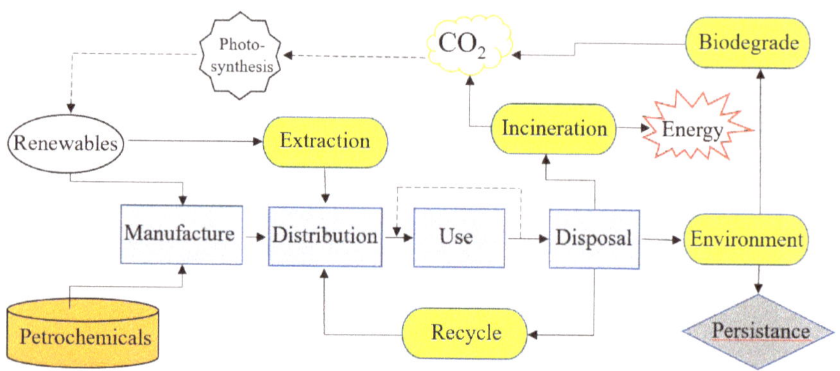

Figure 3 The potential stages in the life cycle of a solvent (adapted from Clark and Tavener 2007).

A solvent can only correctly be termed greener than another when considering the environmental impact that the solvent makes during its production, use and disposal (biorenewable for instance refers only to production). To achieve 'near-total "greenness" of chemical processes, it is important to focus on every aspect of the chemical reaction' (Doble and Kruthiventi 2007). Given the influence of each process step, the Life Cycle Assessment (LCA) method can generate a 'green' calculation. LCAs can be conducted by a variety of individuals, from researchers to large corporate industry leaders, and are used as a technique to holistically assess environmental impacts associated with a product's lifespan, from initial material processing through to the later stages of manufacture, disposal and potential recycling (Fig. 3). By assessing the product or process in its entirety, there is a clearer opportunity to help inform and allow decision-makers to select the process or product which will result in the least environmental impact and the most appropriate selection to improve optimization of the product or process (Loubet *et al.* 2017)

Specific applications and future research

A holistic assessment of whether or not a solvent is greener thus also depends critically on its specific use: does it have a greener application than another under a certain set of circumstances? Studies into LCAs with regard to our solvent use in conservation are therefore helpful and an introduction to LCAs in general, and specifically in conservation, is examined further in Chapter 1.

While neoteric (modern or recently developed) greener solvents may be comparatively appropriate for specific applications in the wider world, in-depth research by cultural heritage scientists and careful testing of these solvents by conservators is of critical importantance for their use within conservation. Although many of these solvents may have been termed 'green', based for instance on their manufacturing process, they may not pose a greener solution since not all traditionally used solvents – both within the industry at large and conservation specifically – require replacement. Using a novel modern solvent as a direct replacement for a more traditional solvent may also raise another series of concerns. Assuming good toxicity ratings, more systematic testing must be conducted by conservators and cultural heritage scientists to ensure that these replacements are also safe for their applications.

Towards the ideal

Generally, a greener solvent approach would be extremely advantageous, potentially providing a wealth of improvements including:

- A reduction in environmental and air pollutants.
- Derivatives derived from non-food biomass, which would open up new markets for those operating in the agricultural sector.
- Recyclable solvents and processes, which would present both environmental and economic benefits.
- Safer materials for conservators to work with regularly thereby improving health and safety conditions in workspaces.

Ideally, a greener solvent will generally have a lower human toxicity level and would promote a safer environment in which to work. While some of the further research required is discussed in more detail in Chapter 3, those impatient to embark on greener solvent use need not delay, as this chapter also provides a step-by-step guide to introducing greener solvent solutions into practice immediately.

With the widely different applications of solvents used in conservation on a potentially complex variety of original materials, it is recognized that the selection of a specific solvent by a conservator often requires a high level of expertise and is dependent on many factors. The aim of this handbook is to enable an additional consideration of greenness within this procedure for the benefit of all.

1

Considerations in defining greener solvents for conservation

Karoline Sofie Hennum and Marie Desrochers

Introduction

With climate change a matter that concerns everyone, conservators are moving towards more environmentally sustainable practices with the aim of reducing their carbon footprint by reconsidering their energy and resource use, as well as their production of waste (de Silva and Henderson 2011). Regarding solvent use in general, there is no truly 'green' solvent: 'green' is a term that has been applied to the scale of impact any solvent may have on the environment. The concept of 'greener' solvents is still relatively new within conservation, and there is a need for more specific research. In the meantime, the field has generally relied on theories and practices from green chemistry and the pharmaceutical industry when considering the impact of solvents used in conservation practice.

In defining 'green' developments in the chemical and pharmaceutical industries, researchers have aimed to create comprehensive frameworks in which solvents are evaluated on their environmental impact as well as on their health and safety implications. There are assessment approaches that use Environmental Health and Safety (EHS or SHE) data on the inherent hazards of a solvent such as in Capello *et al.* (2007). Life Cycle Assessment (LCA) meanwhile is a risk-based approach that quantifies impacts based on potential for exposure. The environmental impact of a given solvent or material is evaluated by calculating the emissions associated with its production, use and disposal. These two approaches (risk and hazard) seek to map the complexity of any single material's impact on both environmental and human health over time – a monumental task. In addition to summarizing these methods, this chapter introduces the STiCH project, the CHEM21 selection guide, and the Solvent Star, preceded by a classification of the solvents generally encountered in conservation and their typical applications.

Classifying the solvents used in conservation

In the field of conservation, solvents, both inorganic and organic, are among the most commonly used materials by conservators. Water is an example of an inorganic solvent while organic solvents are based on carbon and classified by their chemical structure. Conservators from every specialism most commonly use solvents for treatments involving cleaning and for creating solution-based adhesives and coatings. When faced with the cleaning of surfaces on a range of objects, conservators may use solvents to dissolve and remove dirt and coatings of an organic nature such as grease, varnishes and paints (Conservation Unit Museums and Galleries Commission 1992: 61). Adhesives created by adding solvents are called 'solution adhesives'; these are produced by dissolving an adhesive material in a suitable solvent. By adding an adhesive such as this to a surface, the solvent will evaporate, leaving behind the solid adhesive.

Aqueous solutions

This handbook is concerned primarily with organic solvents, thus aqueous solvents will only be discussed briefly. Water is the most obvious 'greener' solvent and can be used to create solutions for some conservation material applications. For instance, in creating consolidants for some organic and archaeological materials, polyethylene glycol dissolved in water is used to consolidate wet organic materials before drying (Caple 2000: 112–18). For some applications in conservation, aqueous solutions may sometimes be preferable to organic solvents as they can be much less hazardous to human health, depending on the additives used. They are generally used to remove soiling material that has become attached to original surfaces or intermediary coatings. To many, soiling removal is more commonly known as surface cleaning (Stoner and Rushfield 2012: 501). Within some disciplines, such as textile conservation, it is rare for a washing solution intended for a cleaning treatment to consist of water alone. Instead, we often find dispersions and emulsions. Such emulsions often consist of distilled water in addition to cleaning solvents and surfactants (Tímár-Balázsy and Eastop 1998: 203–65). Another frequently used, and readily obtainable, aqueous mixture utilized in conservation, especially for cleaning painted and some other surfaces, is saliva (Banti *et al.* 2019: 451–61). There are many other aqueous solutions and mixtures that are important in conservation, which might contain chelators, detergents, surfactants, gelling and thickening agents, and acids and bases. However, it is worth noting that aqueous solutions are not always the greener choice: considerable energy is required in the production of purified or deionized water for example, and disposal of these aqueous mixtures may have negative downstream environmental effects. All these factors are calculated in the creation of a Life Cycle Assessment.

Organic solvents

Aqueous solutions are also not always effective or appropriate for the desired purpose, and a conservator may preferentially choose an organic solvent to create an adhesive, a cleaning solution, coating or consolidant. Organic solvents can be defined according to the three main different structure types:

- hydrocarbon solvents (aliphatic and aromatic)
- oxygenated solvents (alcohols, ketones, aldehydes, glycol ethers, esters, glycol ether esters)
- halogenated solvents (chlorinated and brominated hydrocarbons).

One way to consider them is according to their polarity. Hydrocarbons (containing hydrogen and carbon only) tend to be less polar while alcohols, ketones, ethers, esters are more polar (Conservation Unit Museums and Galleries Commission 1992: 62–71): all are less polar than water.

Organic solvents can be further characterized as saturated/unsaturated, linear/cyclic and aromatic/aliphatic. Within these subgroups, solvents can fall under two or more classifications. For instance, alkanes such as methane, ethane, propane and methylpropane (isobutane) are saturated hydrocarbons. In the group of unsaturated hydrocarbons, we find alkenes, such as ethene (ethylene), propene (propylene) and butadiene. Within both of these subgroups there are cyclic hydrocarbons (either saturated or unsaturated), for example, cyclopropane, cyclohexane (saturated) cyclohexene and cyclohexyne, cyclohexa-1,4-diene and cyclohexa-1,3-diene (unsaturated). Aromatic hydrocarbons such as toluene, benzene and xylene are unsaturated, cyclic and aromatic. They contain the benzene ring (which presents a particular delocalized type of bonding) and have an associated polarity. Many aromatic hydrocarbons are known to be extremely harmful, both to the environment and humans (Mills and White 1994: 1–6).

Assessment methods

Environmental, Health and Safety (EHS)

In the EHS assessment approach, nine categories are evaluated including fire and explosion potential, acute and chronic human health impact, and toxicity for water and air. A chemical is given a numerical index value for each category, also between zero and nine, which determines the overall 'EHS indicator score' (Capello *et al.* 2007). The method, which may also be applied to mixtures, allows for hazard communication in graphical representations such as bar graphs that can plot all categories at once for ease of comparison. This method assesses the properties of solvents at the point of use, collating data from a variety of sources citing health hazards. It complements the Life Cycle Assessment (see below) method by applying a human health perspective to quantified hazards present in any given solvent or mixture.

Life Cycle Assessment

The Life Cycle Assessment (LCA) method considers all stages of a solvent's life, all aspects of its manufacture, transport, methods of use, and eventual disposal. In this way it comprises both the direct and indirect environmental impacts of any solvents in a given process. To effectively capture all the complexity inherent to this method, software is employed to calculate the final value used to compare one solvent to another. These two methods may be used together: EHS-based methods assess inherent hazards of the solvent itself, while the LCA method assesses potential environmental impacts across the entire solvent life cycle.

Life Cycle Assessment (LCA) is a popular, internationally standardized modelling tool used to quantify the environmental sustainability of a material or activity over its technological life cycle, from its manufacture, through its use and eventual disposal (ISO 14040:2006). It can be used to track overall resource requirements such as energy or water, emissions and the environmental and health impacts caused by these emissions. The pre-consumer life cycle, defined as cradle-to-gate, encompasses inputs and emissions covering the extraction of resources, processing, and assembly, up to the point that a saleable product or material crosses the factory gate (Patel 1999) (Fig. 1). A more expansive cradle-to-grave LCA also includes inputs and emissions from product use and end-of-life (reuse, recycling, waste treatment and disposal). Considering the entire product life cycle is critical to understanding all the direct and indirect ways in which a product affects the environment: often the greatest drivers of emissions are hidden from consumers, either upstream in the supply chain or downstream in waste management. A related concept is cradle-to-cradle, which is a design philosophy that aims to shift the current linear system of once-only manufacture, use and disposal to a circular approach whereby durable materials are recovered and recycled while other materials are made from benign, bio-based sources and can be safely returned to nature.

LCA has been used to evaluate thousands of materials and products and is in common use across all sectors, including, increasingly, in the cultural heritage field. It is employed by businesses to evaluate the products they manufacture and to find ways to improve their environmental performance; by consumers to understand the global implications of the products they buy and to choose among options; and by policy-makers to set system-wide emissions and environmental targets for society. Consumers and clients are increasingly calling

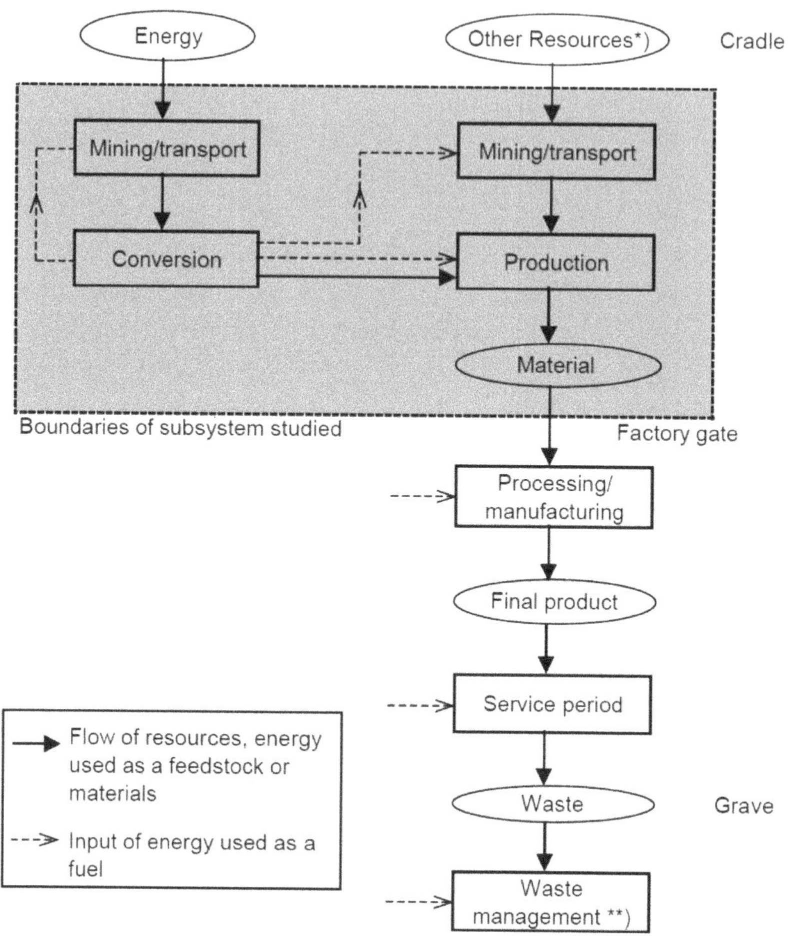

Figure 1 Overview of the entire system and subsystem. Closing Carbon Cycles: carbon use for materials in the context of resource efficiency and climate change (Utrecht University Repository, diagram from Patel 1999). *Energy analyses do not account for the consumption of resources other than energy but they do include the energy requirements to process these 'other resources'. ** Landfilling, incineration and recycling (not shown in this figure).

for greater transparency with respect to product supply chains. In response, manufacturers might publish an Environmental Product Declaration (EPD), which summarizes the LCA results for an individual product. For chemicals specifically, LCA can be used in concert with green chemistry principles where human toxicity levels and environmental effects are key aspects for manufacturers to consider when redesigning chemicals for overall safer use (Anastas and Warner 1998).

There are numerous ways in which emissions can cause environmental and health impacts. LCA is a multi-criteria tool that can simultaneously quantify impacts to resource scarcity, global atmospheric change, local air and water quality, and chemical toxicity to ecosystems and humans, among other categories. Several popular footprinting tools fall under the umbrella of LCA, including assessment of embodied energy or embodied water to understand the overall resources needed to produce a material or product, or embodied carbon that tracks the carbon footprint or life cycle greenhouse gas (GHG) emissions and their contribution to global climate change. In conservation, LCA has been used to investigate a range of environmental questions, including carbon footprinting of loans, comparison of treatment and preservation methods, even sustainable design of displays and exhibitions.

In July 2021, Sustainability Tools in Cultural Heritage Preservation (STiCH) launched the LCA tool and case study library, a valuable resource for the conservation community when making material decisions. STiCH provides an open access platform for comparing storage, transport and conservation treatment materials using their carbon calculator tool (https://stich.culturalheritage.org/) (Fig. 2). The library includes case studies comparing the carbon footprint and environmental impacts of papers, textiles, solvents, adhesives and other materials used regularly in conservation. The tool allows users to rapidly calculate and compare cradle-to-gate emissions, GHG emissions for each material and product option, while including links to Safety Data Sheets (SDSs) for health and safety and environmental impact toxicity concerns. Case studies present full LCA results of more complex environmental comparisons, such as which type of crate system is environmentally preferable. The STiCH project has been supported by grants from the National Endowment of the Humanities in partnership with the Foundation for the American Institute of Conservation (FAIC) and Northeastern University (Boston, MA).

Justine Wuebold and Matthew Eckelman

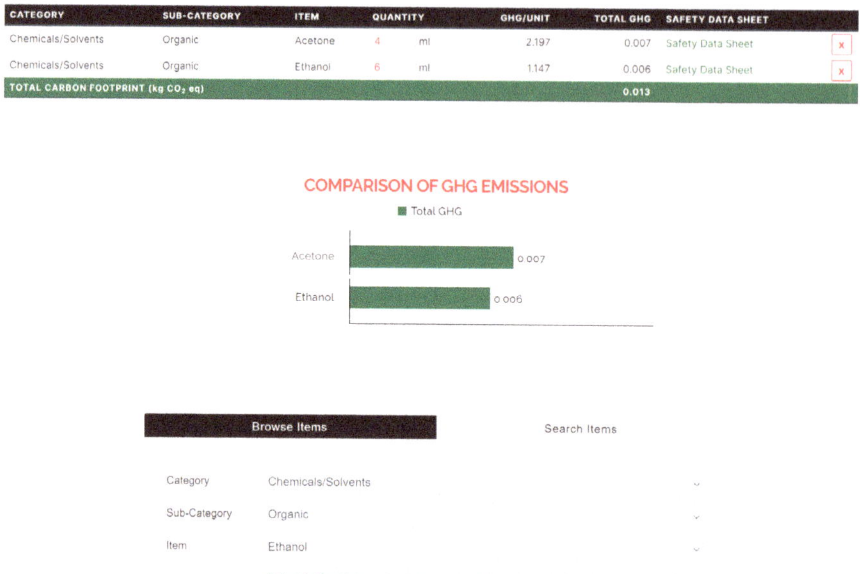

Figure 2 Arbitrary sample of comparison between acetone and ethanol in the STiCH carbon calculator tool illustrating how the tool works and the categories available.

Greener approaches and the ideal solvent

The pursuit of greener conservation work requires each of us to think holistically about the processes we have been trained to utilize and the ways in which we work each day. Acknowledging that a solvent is not used in isolation, that it may be selected and used in a highly technical and specific way, requires us to consider the entire conservation process involving its use. Therefore, prior to solvent selection, options for entirely new ways of working and adapting traditional practices to fit within the EHS and LCA evaluation frameworks can be considered. If solvent use is required, all aspects of the solvent's life cycle and the conservator's practice must be considered in determining, within the best treatment option for the object, the least impactful approach.

A greener solvent for our applications may mean preferential selection of one that is lower in toxicity for human health and impact on the natural environment. Thinking of the solvent in terms of each stage of its life – from its production, transport, application and disposal – a greener solvent will also require less energy usage at each step, and overall its disposal will produce less harmful waste and environmental pollution. Additionally, efficacy and efficiency in use is also considered. A solvent that requires less volume to perform a task may be preferable to a less efficient solvent that requires more volume for the same application.

A note on differing perspectives for safety considerations: European vs. American SDSs

International chemical safety regulations and labelling standards have implications on the accessibility of hazard information as well as the availability of certain solvents from one country to another. In 2007, the European Union adopted the Registration, Evaluation, Authorisation and Restriction of Chemicals (REACH) regulations, changing the way consumers and manufacturers interact with and produce chemical products. These guidelines set standards for marketing any chemicals within the EU across all sectors. It focuses on transparency and protecting consumer health as well as the environment, and places the onus of testing and hazard communication on the manufacture (European Chemicals Agency 2021).

Since 2003, the United Nations' Globally Harmonized System of Classification and Labelling of Chemicals (GHS) has become a standard for hazard communication and chemical labelling in the form of chemical safety data sheets (SDSs) (Fig. 3). This method has been adopted by nations worldwide despite differences in chemical safety regulations and

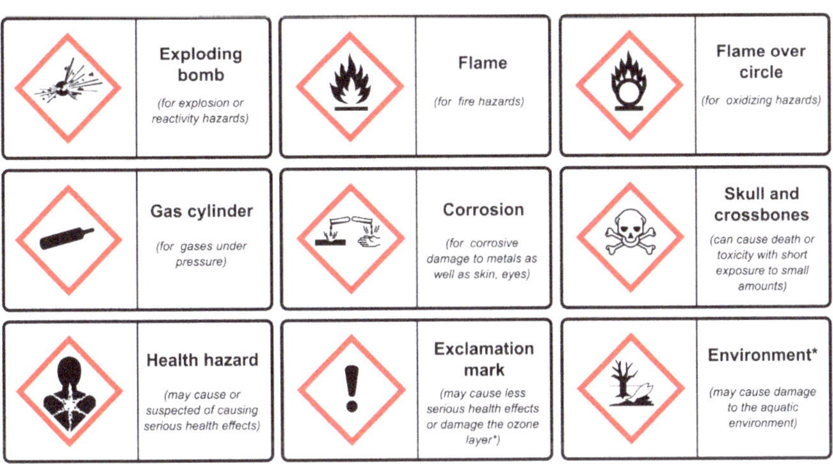

Figure 3 Pictograms for hazard communication using the Globally Harmonized System.

communication policies. This handbook focuses on the hazard communication and policies of the European Union because it has the strongest regulations.

Decision-making

When deciding on which solvent to use for a conservation treatment, a conservator must consider a range of factors, the most important being to firstly consider whether or not a solvent is really needed for the chosen conservation application. If a solvent is required, the conservator should then consider which solvent option would be the greenest taking into consideration how long the solvent will be used and how much is needed. However, it may be extremely difficult to determine this based simply on an individual's experiences and knowledge, in which case other assessment resources may prove very helpful.

In 2016, the CHEM21 selection guide of classical and less classical solvents for the pharmaceutical industry was published (discussed further in Chapter 3). This selection guide contains many common solvents, including the conservation-relevant water, alcohols, ketones, esters, ethers, hydrocarbons and acids. A methodology for considering factors such as exposure values, health and environment criteria, solvents have been accorded scores resulting in a ranking category: '**Recommended**', '**Problematic**', '**Hazardous**' and '**HH**' (extremely hazardous). Conservators can use this selection guide as a tool when deciding on which solvent to use, preferably together with other tools and considerations of solubility data as suggested in 'The Solvent Star: assessing and documenting solvent selection' (Fife 2020) (Fig. 4). A methodology for using some of these decision-making tools to replace the most harmful and hazardous chemicals with greener options is provided in Chapter 3.

Conclusions

It is important to clarify that many different factors and tools can aid conservators to assess and determine the greenness of a solvent they are planning to use. It is not only the properties of the solvent itself that functions as a deciding factor, but also the greenness behind its production

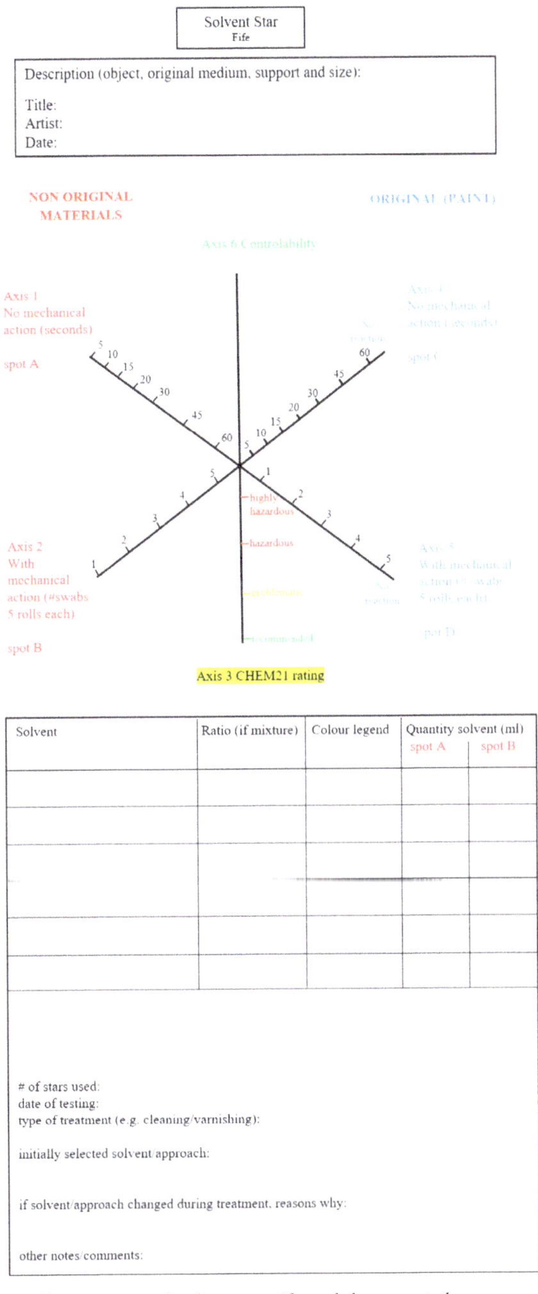

Solvent Star
Fife

Description (object, original medium, support and size):

Title:
Artist:
Date:

Solvent	Ratio (if mixture)	Colour legend	Quantity solvent (ml)	
			spot A	spot B

# of stars used:
date of testing:
type of treatment (e.g. cleaning/varnishing):

initially selected solvent approach:

if solvent/approach changed during treatment, reasons why:

other notes/comments:

Figure 4 The Solvent Star was conceived to quantify and document the conservator's small solvent tests and thereby assist in determining the most safe and efficient solvent for both practitioner and artwork in any individual application. In addition to the quantitative testing axes (1, 2, 4, 5) there is an axis denoting the CHEM21 rating of the tested solvents (3) and a qualitative axis (6) for practitioners to express their feelings/comfort with the solvent use/action (adapted from Fife 2020).

processes, the health impacts of human exposure, and finally its effects on the environment. Since the measurement of the greenness of any given solvent is relative for any specific application, all the available options must be considered taking into account all these factors. This movement and system thinking within conservation is gaining momentum. Tools such as STiCH's Carbon Calculator and Fife's Solvent Star can make a difference to decision-making by helping practitioners to sort through all the information available. Creating a culture of thoughtful decision-making, based on the universal desire for greener solutions, is a first step to supporting one another in this journey.

2

A history of 'green' solvents in conservation

Aline Assumpção and Lucile Pourret

Solvents used in conservation through the ages

Solvents play an essential role in many conservation treatments whose existence long predates the well-established norms and modern conservation products. Since the field was first established, conservation and restoration has been plagued by many uncertainties due to the lack of documentation and the lack of attention paid to it at the time. Conservation evolved alongside science, society and the history of art. In Europe, the restoration of paintings has records of its practice since the 13th century. Although they might not have been based on today's ethics and protocols, chemistry has always played an important role in the processes.

Washing, remaking, refreshing, and lightening were terms found to describe the work of paintings preservation in the 15th century (Conti 2007). It was often carried out by artists, and the work could involve repainting, over-painting, excessive cleaning, and an exaggerated completion of missing parts.

The term restoration began to be used more widely in the 16th century, but was still carried out as an empirical practice, with a sense of remaking, to render the artwork looking good and new. Recipes for cleaning could include ashes, lye, urine, lime, honey, soaps, vinegar, eggs, salts, alkalis, acids, water, saliva, bile fluids, several sliced foods, alkali, potash, essential oils, butter, olive oil, amongst others (Conti 2007; Hubbard 1775; De Piles 1776).

Polarity and solubility concepts were applied in an empirical manner. However, the role of chemical theory played an increasing part in the development of conservation, as can be deduced from the fact that many of the authors of historical restoration recipes were chemists, apothecaries, and scientists (Stols-Witlox 2019). The need to understand what we are dissolving or affecting had already been mentioned in 1758, that is, the need to remove only selected layers without damaging the substrate we wish to preserve. However, a more selective cleaning methodology was not well diffused before the 19th century. For the removal of oxidized varnishes, solvents could of course only be recommended from among those available at the time, sometimes resulting in excessive cleaning and the removal of glazes. Montabert (1829) advised readers to first attempt dry cleaning or use pure water before proceeding to (organic) solvents or alkalis. If it should prove necessary, they would increase concentration or polarity, prolong exposure time or add other combined materials (Stols-Witlox 2011). It was gradually established that different areas of the surface may have different sensitivities and that all cleaning action has consequences for the artwork.

Stols-Witlox describes solubility tests as starting with turpentine oil, sometimes mixed with a drying oil. When historically considered 'stronger' solvents were required, 'ethanol was employed (white wine, wine spirit, brandy, alcohol of different purities), added gradually to turpentine oil until the desired effect was obtainable' (Stols-Witlox 2011: 4). Essential oils were mentioned for cleaning with warnings regarding their aggressiveness; ether was suggested for the removal of overpaint or oil layers and acetone mixed with water to remove resistant varnishes (Vibert 1891).

Organic solvents – although lacking the same purity and quality of today – have been used for centuries in conservation, sometimes in mixtures such as alcohol with essential oils and turpentine. Experimental chemical production of organic solvents began in the late 18th century and developed rapidly throughout the 19th century (Spurgeon 2006). At the end of the 19th and beginning of the 20th century, with greater knowledge of solvents and polymers, there was a growing dialogue between science and restoration concerning the possibility of their application in treatments. In the 19th century, Pettenkofer introduced a process using alcohol and copaiba balsam to treat blanched surface coatings. Gradually, restoration treatments became safer, laboratories started to be installed in museums, and the international ethical norms for cultural heritage preservation began to be established.

The petrochemical industry evolved throughout the 20th century to become ubiquitous for the production of cosmetics, paints, glues, detergents, plastics, among many other everyday products, and by the second half of the century, society was already largely dependent on them. This dependence also affected conservators: conservation science grew exponentially as new and purer products became available. Greater quality control and accessibility, along with solubility efficacy, made organic solvents popular within conservation-restoration treatments.

Organic solvents are carbon-based solvents with properties that differ according to their structures. They can also be divided into natural and synthetic classes: naturals are those produced by naturally occurring biological processes such as ethanol while the synthetics require the chemical reaction of other products, such as xylene, in order to be produced. Most synthetic organic solvents are currently derived from petrochemicals.

The potentially neurotoxic effects of organic solvents have been well recognized since at least the 19th century (Andrews and Snyder 1986), and possibly earlier. They were then further studied by occupational medicine (Spurgeon 2006). These findings also contributed to the growing environmentalism that took place at the end of the 20th century. These hazards are now widely reported: depending on the type/extent of exposure, it is known that even the least toxic solvents may have a harmful effect on health. Moreover, these solvents impact not only professionals working in direct contact with them, but also all living beings that are affected by contamination of the soil, water, air, etc. (Joshi and Adhikari 2019). At the same time, modern organic chemistry requires organic solvents and society today relies on them for many different purposes.

Developments in considering the health and environmental impact of solvents in conservation

The history of solvents used in conservation and restoration has naturally followed developments within the chemical industry as a whole, in particular the petrochemical industry. Since the 1970s, a broader awareness of the potentially negative impacts of certain chemical substances on aspects of the surrounding natural world has become well established. This understanding goes hand-in-hand with increased knowledge regarding the specific interactions and the heightened effect from the expanded use of such materials according to elevated societal demands (on larger scales with larger quantities). The conservation field has also been impacted by these changes.

As previously mentioned, most of the organic solvents used in the field of conservation and restoration are created from industrial petrochemical processes which started in the 1920s and further developed after the Second World War. Legislation to regulate these solvents was swiftly introduced with a view to marketing them as goods. Taking the example of Europe, in 1948, the former EEC issued a first regulation on solvents, called CLP, with the aim of regulating the labelling, classification and packaging of chemical products. A modification of this directive was made later in 1967, which in 2015 was refined to become the directive currently applied. However, this directive does not provide more precise information on the toxicity of these chemicals for humans or for the ecosystems in contact with these substances. It was also not the principal aim of this directive. According to a protocol of 1991, Europe legislates to fight against the emissions of several atmospheric pollutants and their transboundary flows. In 1999, Europe legislated on volatile organic compounds (VOCs) to reduce their use in industry and finished products such as industrial paint. But it was not until the REACH directive, which came into force in 2007, that the issue of protecting health and the environment against the risks associated with chemical substances became a specific focus (Fig. 1).

With respect to considerations on the health of humans, such as conservator-restorers exposed to or handling solvents, awareness is now quite progressive. By the second half of the 19th century, disorders related to the handling of solvents were already being noted. The French chemist

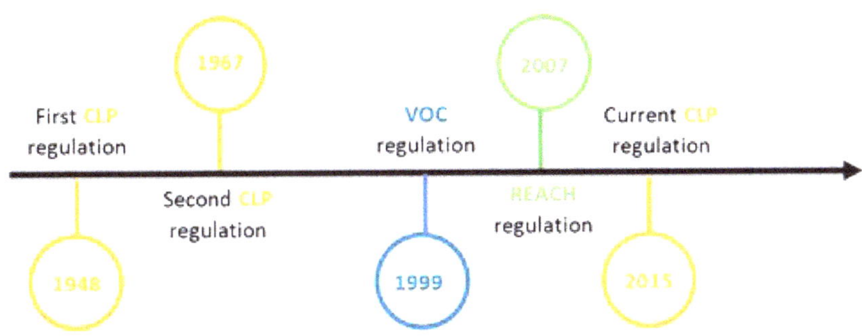

Figure 1 Timeline of regulations.

and industrialist, Anselme Payen, was concerned about the use of carbon sulfide in closed environments. This observation, supported by other studies in the rubber industry, led to the installation of ventilation systems in factories at the beginning of the 20th century.

In the early 1960s, more precise studies were carried out which linked anxiety, depression, loss of memory and concentration symptoms of certain workers in the viscose industry in Finland having been exposed to small doses of carbon sulfide over many years. The book *Silent Spring* published by the American biologist Rachel Carson in 1962 proposed the theory that exposure to small doses of many chemical substances used in industry and agriculture over a long period of time can lead to health problems years later, and that the effect of solvents is cumulative. Research in the years following this revelation was to prove the validity of this claim.

However, it is in the second half of the 20th century that an acceleration of the awareness of the detrimental effect on the health of workers exposed to chemical substances can be observed. More than 40 studies conducted between 1970 and 1990 on the neurological effects of exposure to organic solvents have removed any residual doubts as to the occurrence of the psycho-organic syndrome (POS) caused by solvents. At the same time, in 1968, the Institut national de recherche et de sécurité (INRS), the French National Research and Safety Institute for the Prevention of Occupational Accidents and Diseases, developed applied research, initially focused on the collection and dissemination of information. Since then, it has conducted research into the toxicity of various solvents for the human organism in the working environment.

Even though these studies are more often conducted by and for the chemical industry, occupational medicine and researchers are also interested in conservator-restorers who are in regular contact with organic solvents. An article published in 2003 by the INRS concerned the toxicological risks to which conservator-restorers are exposed. The article mentions in particular the lack of study carried out to date within this profession and the difficulty of establishing precise results due to an inadequate population size for the statistical sample. In 1996 another study on the same subject was published (Tomei *et al.* 1996).

As the use of organic solvents has become more extensive, new problems have appeared. Between the 1970s and 90s, more scientific and statistical studies were conducted aimed at regulating the use of organic solvents harmful to humans and ecosystems, leading to increasingly precise legislation. Laws regulate the use of organic solvents and some countries, in the European Union for example, even ban the use of certain solvents considered too harmful for both humans and the environment.

Mitigating against the harmful effects of organic solvents

Following the awareness of these problems, solutions were sought to try to protect users and their environment. Since the 1990s, conservator-restorers have invested in more effective means of protection such as personal protective equipment (PPE) including filtering masks for solvent vapours and protective gloves (Fig. 2). Conservator-restorers have also employed practical ways to distance themselves from solvents. For instance. they traditionally use wadded sticks to apply the solvent for varnish removal and now, in addition to masks and gloves, wear Tyvek suits when applying varnish to limit exposure and avoid deposits on the skin and clothing. Studios are sometimes also equipped with suction hoods with filters that capture the solvent vapours preventing their release into the surroundings. This is a protective measure for both the practitioner and the environment. The use of these filters requires their maintenance, as well as their replacement

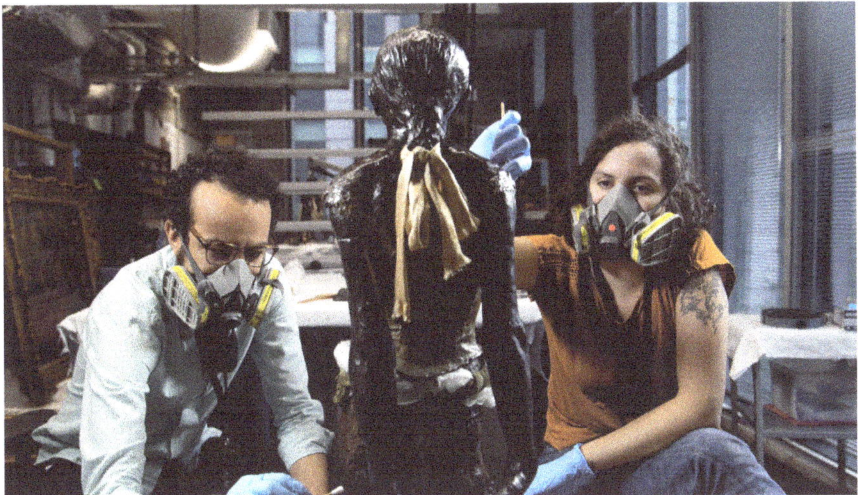

Figure 2 An example of conservators' typical working conditions with selected personal protective equipment (PPE) during the cleaning process of *The Little Fourteen-Year-Old Dancer* by Edgar Degas, 1880, collection Museu de Arte de São Paulo Assis Chateaubriand. Photo: Leno Taborda/MASP.

and management as chemical waste at the time of their obsolescence (or their recycling if this is possible). This often necessitates the use of external service providers.

As professionals using chemical substances, conservator-restorers are obliged to use them according to the handling, storage and collection advice given on the product data sheets and to obtain this information from the manufacturers or suppliers. Many different countries have laws that also regulate the use of chemicals in the workplace to protect the health of the user and the environment. For example, in the French Labour Code and in the UK's Control of Substances Hazardous to Health Regulations (COSHH) 2002, if exposure cannot be adequately controlled in any other way, workers should wear PPE and be trained in its proper use and to understand its limitations.

Although the implementation of these regulations remains heterogeneous in the conservation-restoration field as professionals sometimes encounter practical difficulties, these regulations are nevertheless very important from the point of view of the health of professionals (short- and long-term protection), working comfort and environmental impact. They must be followed and adopted in daily practice. This also applies to the correct waste disposal of solvents or any product containing chemical substances (e.g. cotton used for removing varnish from paintings). They are treated either by combustion or by recycling (possible for certain organic solvents such as ethanol via a distillation process that separates the pure solvent from contaminants and impurities).

The EU Registration, Evaluation, Authorization and Restriction of Chemicals regulations (EUR-Lex 2006) and the directive regulating the emission of VOCs (1999/13/BCE) replaced by the directive on industrial emissions (integrated pollution prevention and control) (2010/75/UE) classify solvents according to their toxicity for the environment and lists those must be disposed of by specialist companies. In Spain, the law compels all professionals who use solvents to hire companies to manage this waste with controls in place to ensure compliance. In order to be more sustainable, it is essential that conservator-restorers have appropriate containers for used solvents and chemical waste that cannot be disposed of with the normal rubbish. More and more contractors are offering their services to small businesses and craftsmen who work alone and do not have a large annual quantity of solvents to process. It is usual to dispose of solvents in several containers: water-based solvents together, with halogenated solvents (in conservation-restoration most often chlorinated solvents) and the rest (hydrocarbon and oxygenated solvents) in separate containers. However, there may be other requirements, such as the separation of acids and bases, which can also be treated differently depending on the contractor. The processes of waste management and protective equipment, although very practical in reducing the impact of solvents, can be restrictive both in their implementation (e.g. workshop space) and daily practice (e.g. advised maximum periods for wearing a solvent mask). Thus, using less solvents and chemical products must remain a primary aim in reducing their impact.

Since the end of the 1990s, several studies in the conservation-restoration field have been heading in this direction, primarily by proposing alternatives to replace teratogenic, carcinogenic solvents by other, less toxic ones, as in the case of benzene which was replaced by toluene in the 1970s in most countries. In French workshops, the use of benzene as a thinner was prohibited by law following a decree dated 14 June 1969 as it had been found to be highly carcinogenic.

In the search for alternatives, a symposium was held in France in 2010 (Palmade-Ledantic and Picot 2010) concerning conservation-restoration and the safety of practitioners working in the field. The purpose was to reflect on the replacement of the most toxic solvents by less toxic alternatives. Several solutions were discussed, including the use of the Teas chart (discussed in more detail in Chapter 3). To reduce the amount of solvents used in their studios, some conservator-restorers have employed certain thickeners, gelling agents, and other ways of adapting their solvent application methods. Such adaptations can reduce the quantity of solvents used and slow down their evaporation.

Towards a greener future

The use of solvents poses several problems that have been identified over the years, often with corresponding attempts to resolve them. As described above, solutions are constantly evolving thanks to a clear acceleration since the 1990s of research in this area. Understanding the impact of solvent use is reaching a wider audience. Over the last decade, the conservation-restoration community has initiated many changes in its practices regarding the use of solvents.

Research is increasingly being carried out in specialized chemistry departments, cultural institutions, and centres developed for the study of cultural property and new restoration techniques. According to the results of the 2020 survey on green(er) solvents mentioned in the Introduction, the profession at large is continually seeking information to enable improvement in conservation practice by utilizing more responsible and thoughtful solutions. Faced with this growing demand, new products are appearing on the market labelled as 'green' or 'eco-responsible'. Frequently of different natural or petrochemical origins, and often marketed as 'environmentally friendly' and 'non-toxic for humans', the lack of comparative information, experience and research may make it difficult to apply these products within treatments in studios today. Taking the interesting example of toluene, previously (and relatively speaking, correctly) introduced as a safer alternative to benzene, toluene became widely used in studios. Now, however, toluene is considered a relatively harmful choice for practitioners and ecosystems for which we must seek safer alternatives. Although our reliance on organic solvents for conservation treatments has been apparent since treatment recipes were first recorded, scientific awareness within the profession alongside green chemistry principles has led to the pursuit to reduce this dependence. In the first instance this is demonstrated by methods to reduce the amount of solvents used in processes and to optimize material properties.

The improvement of aqueous methods, viscosity modifiers, some branches of nanotechnology, micelle science and surfactants is of importance in this research and, although not without any concerns regarding potential impacts, can provide alternatives to more harmful solvents used in some cleaning applications. Water, for example, has been used since antiquity in the removal of soiling materials on painted surfaces, and conservation science has shown that adjustments can render it suitable for removing coatings or other complex materials (Wolbers 2000).

However, potential contamination of the environment with waste products and the processing requirements of production and recycling when using water-based systems should also be carefully considered. Clean (pure) water should not be viewed as an unlimited resource, and the energy consumption from the production of deionized and distilled water should be considered. The further high energy costs of water recycling and the impact of additives present in modified aqueous systems – such as surfactants, chelating agents, polyacrylic acids, etc. – and their proper waste disposal should be determined and implemented. While aqueous solutions will generally present the greener option, it remains possible that, for a particular application, use of a more efficient and technically appropriate organic solvent with good SHE ratings and a low LCA score may have a lower environmental impact.

New technologies are essential to chemistry and society's development, and organic solvents play a necessary role and possess desirable properties in many applications. However, existing alternative organic solvents or mixtures with lower toxicity and lower environmental impact should be considered for replacing more harmful solvents which have been routinely used for preparing adhesives, varnishes and for coating/soiling removal (see Chapter 3). The impact on the original substrate is one of the key reasons for these choices and must be addressed to prevent damage to objects and to provide a more complete 'picture' of the argument for change.

As complex as the subject already is, disparities in realities around the globe regarding legislation, accessibility and information make for a very uneven playing field. In countries outside Europe and North America, the evolution of conservation practice has proceeded somewhat differently, with a general time-lag. Besides these different timelines concerning conservation materials, it is important to bear in mind that each country has different regulations concerning chemical products, so accessibility is dependent upon locality. Moreover, an important aspect to sustainability involves the fabrication and transportation process, so using locally sourced materials is always a factor to be considered when evaluating the 'greenness' of the materials we use. As was seen in Chapter 1, these aspects and impact evaluation are a necessary inclusion in the LCA process.

3

Practical steps to greener solutions

Gwendoline R. Fife

Introduction

The premise of this chapter is a 'how-to' practical guide to finding greener solutions in conservation treatment situations where it has been determined that the best (or only) option is to use organic solvents. As a paintings conservator and teacher on these topics, I have often found it striking that when solvents are mentioned in conservation-related contexts, 'cleaning' generally follows. This is despite the fact that methodologies developed over the last 40 years have given conservators numerous options to reduce, and sometimes eradicate, organic solvent use for many cleaning treatments. Whereas the common practice of applying adhesives (varnishes and sometimes consolidants) in the form of reliable and tested resins is one that often unavoidably requires, at least for the present, organic solvents. Therefore, alongside considering greener solvent solutions for removing adhesives and coatings from surfaces (for which they will sometimes remain the most safely efficient option), it is for the application of varnishes and coatings that there is a critical need to incorporate greener solvent solutions into our recipes. This is especially the case when relatively harmful solvents, both for us and the environment, are otherwise included in the mix. Hence, while acknowledging that there are certain complications to be considered, it is the intention of this chapter to show that one of the primary aims of the 'Sustainability in Conservation Greener Solvents' project – *removing the most hazardous solvents from our conservation organic solvent cupboards*

– can be carried out by following relatively straightforward steps. The approach outlined below aligns with both long-standing practice in many conservation studios worldwide (Hedley 1980; Burke 1984; Pietsch 2002; Rivers and Umney 2003; Vila and García 2013), as well as the industrially termed Greener Solvent Alternatives approach, wherein greener substitute solvents (or mixtures) are determined to replace more traditional solvents that have health or environmental concerns.

When considering our aim to replace the most harmful solvents and apply a greener solvent approach we can pose three main questions:

- What do we really need to replace?
- How do we identify a greener replacement?
- How do we safely apply this to our work?

These three key questions underlie the following step-based process for assessing and immediately implementing greener solvent use. Further information regarding the basis for each step is provided in the explanatory notes below.

Greener steps

STEP 1 STUDIO SOLVENT CUPBOARD ASSESSMENT

- Open your studio solvent cupboard.
- Open the **CHEM21** publication link to Table 7: https://pubs.rsc.org/en/content/articlepdf/2016/gc/c5gc01008j (Prat *et al.* 2016: 292).
- Identify each solvent in your cupboard and check it on CHEM21. If it is '**Recommended**' you can move to the next: '**Problematic? Hazardous** or even **'HH'** (highly hazardous)?' Take it out and examine the safety, health and environment number rating (EHS/SHE numbers). If the health (H) number is 5 or above this solvent should be targeted for removal.
- If it is not on the CHEM21 list check the database on REACH (ECHA) (https://echa.europa.eu/information-on-chemicals) and/or the National Library of Medicine (https://pubchem.ncbi.nlm.nih.gov/). Examine hazard classification and labelling (GHS). For solvents not

included or under review assume problematic rating (in accordance with CHEM21 instructions).
- Read the solvent's associated Safety Data Sheet (SDS) for appropriate handling measures and disposal.

When considering solvent selection we have some readily available assessment tools of which CHEM21 is particularly practical in its application. As stated by the authors, 'This survey may be very useful for the quick selection of a solvent, in particular in academic institutions or companies which do not have their own solvent selection guides.' Life Cycle Assessments/analysis (LCAs) specifically reflecting the typical applications of solvents within conservation should be increasingly beneficial (see also Chapter 1). But in accordance with the ambitions of CHEM21 – as was highlighted by its authors – the straightforward provision of the SHE numbers enables an immediate and complementary application outside the pharmaceutical industry. Within conservation studios, CHEM21 allows practitioners to make a further judgement specific to their priorities regarding these aspects. *In this way it can provide clarity as to what should be removed from our solvent cupboards first.*

Thus, having the data to make an explicit selection based on a specific criteria, *what do we really need to replace?* To determine which solvent is the worst – but with valued working properties making it critical to replace – depends in the first instance on the contents of the cupboard. So for the purposes of an example for this chapter, we consider a hypothetical solvent cupboard containing some classical (traditional) solvents and their subsequent presentation on the Teas chart (see Table 1, pp. 42–3, and Fig. 1, p. 47).

In the case of this solvent cupboard, by looking at the overall CHEM21 rankings first in Table 1, followed by the SHE numbers, an immediate decision can be taken to remove toluene. Although toluene was itself introduced as a replacement for the more harmful benzene (a healthy reminder that it is all relative and safety aspects generally improve), it is a solvent preferably avoided due to potentially serious health effects including reproductive toxicity. For reasons related to potential impacts on human health and environment, xylene (a possible subsitute for toluene) will also be deleted from our hypothetical cupboard. The removal of these two solvents corresponds to the healthy desire expressed by many conservators to minimize their exposure to aromatic solvents.

STEP 2 WHAT NOW? SEEKING SUBSTITUTES AND ALTERNATIVES

As in any industry, there are two options – a single greener replacement (substitute solvent) or greener alternative mixture(s). Both options have advantages and disadvantages, which are dependent on the application and individual substrate in question. Single solvent replacement may have certain theoretical advantages, but mixtures may provide the best solution in practice. For both of these approaches, although limited, Hansen solubility parameters (HSP) can aid us greatly. With more information provided in the notes below regarding these issues, it suffices to say here that for modelling dilute solutions, such as the creation of single polymer solutions, HSP are accurate enough as volume-weighted averages. In recent years it has become much easier to generally access these, and typically for our solvents in question their HSP can be found in various references (Horie 2010; Abbot 2013, 2021). Some modern or more recently developed (neoteric) solvents have been mentioned within a conservation context (e.g. Martinez *et al.* 2017; Prati *et al.* 2019). Their published HSP were used here to generate fractional parameters and assign Teas chart positions (see Table 2, pp. 44–5, and Fig. 2, p. 48). Their inclusion is not a recommendation per se. Considering 'a lot of time and energy is wasted looking at solutions that are not relevant' (Abbott 2021), alongside the generally stretched financial resources in conservation, replacement with a neoteric solvent would be most valid if it is already in your cupboard, and/or is a viable direct replacement for a solvent with valued working properties but targeted for removal. A case study for this approach is presented in Option 2 while a methodology for determining greener alternative mixtures is considered in Option 1.

Option 1 Greener alternative mixture

'We have to go greener by being smarter with what we have' (Abbott 2021). Sometimes, mixtures are the only way to achieve a safe/desired result. Depending on your level of resources, interest and time there are various tools at your disposal regardless of how many or few of these you have.

More tool options and theoretical justification can be found in the notes below, but if you are doubtful about its placement within this context, please note that the *Teas chart is only being considered as an empirical tool – it is of no use for modelling solvent effects on art objects and other complex substrates.* But for dissolving conservation materials (e.g. single polymers) to make solutions (varnishes, consolidants etc.) the Teas chart is typically reliable in its predictions (for mixtures containing up to three organic solvents, and excluding water): it is a practical approach that is also very straightforward to use.

To that end, since toluene and xylene are being removed from our hypothetical cupboard (see Table 1, pp. 42–3, and Fig. 1, p. 47), it is very apparent that we will be left with a significantly empty lower right region in the Teas chart. To add to our woes, the least polar solvents in the lower right corner are not CHEM21 '**Recommended**' either but they do represent a preferred choice at this point. It is also worth noting that while some solvents may currently be produced via petrochemical means, there are possibilities for their development from bio-renewable sources (see the Introduction).

In order to find a successful replacement solution, the following points need to be considered:

1 The solvent parameters of the solvent/region of solvent space to be substituted (in this case toluene).
2 The nature of the treatment application.
3 The potential effects, both visible and long term, on original materials.

Illustrated with joined lines between CHEM21 '**Recommended**' classical (traditional) and remaining solvents from Table 1 (the 'Example Solvent Cupboard') are some combinations that come close to our target area for replacing toluene (see Fig. 3, p. 49).

This initial selection for developing appropriate mixtures benefits from an understanding of solvent properties and non-ideal behaviour in mixtures (for more details see the explanatory notes below). Of significant influence in determining what is practical and desirable for the intended treatment application are the relative evaporation rate (RER) and knowing (from personal and cultural heritage science research findings) which particular solvents (chemical type and/or parameters) are more likely to be problematic for the original materials in the object. While returning to some aspects of this in Step 3 with testing, it is more efficient to consider this in advance of any experimental preparation.

Despite these slightly more complicated facets, a method of locating a solvent mixture on the Teas graph is, as demonstrated in Figure 3, very simply achieved by drawing a line between the two selected solvents in the mix, and finding the point on the line closest to the desired target (in our example, toluene). With this selected point directly corresponding to the volume fractions of the mixture, the suggested ratios for the desired mixture is hereby indicated. The fractional parameter of the mixture can then be calculated from the fractional parameters of the individual solvents (as shown in Table 3, p. 46). Using the fractional parameters of a determined binary mixture as the new 'single solvent', adding a third solvent can also be considered. While one of the greatest benefits of this approach is its accessibility, greater solubility prediction accuracy can of course be provided if you have access to calculations with HSP (see the explanatory notes below).

For the uninitiated, this may appear challenging, but with increasing acquaintance, the speed of this step reduces to the point where it may become unnecessary to consult the TEAS chart at all, especially for preparing habitually used resins. The determined solvent mixture for the targeted application (for which toluene was previously used) can be tested by preparing a small volume with the intended resin. If dissolution and handling properties are satisfactory move to Step 3. An alternative to finding mixtures is to look for a single solvent replacement, and for this it is hoped that the continued creation and suggestion of substitute solvents may present increasingly interesting options for the cultural heritage scientist and conservator to consider. Outlined in Option 2 is a case study illustrating the practical perspective for such a potential introduction.

Option 2 An investigative case study into the single solvent option

As a paper conservator, my experience of using solvents in treatments is while cleaning or removing tape stains, modern adhesive and, less frequently, varnish. The second principle of green chemistry is 'Atom economy' (Anastas and Warner 1998). Good atom economy requires most of the atoms of the reactant to be incorporated in the desired products, with the result that only small amounts of unwanted by-products are formed, leading to fewer problems associated with waste disposal. Atom economy is very relevant when removing stains from tape and adhesive. The solvent will dissolve the original sticky element of the stain and reduce the material in the paper but not necessarily remove all of the stain. The stain is composed of many materials such as environmental dirt, aged binder, acidity from cellulose fibres and plasticizers. It is worth considering removing the material in several steps, i.e. using the solvent at the right time, therefore using less of it. Some material may be able to be removed mechanically when dry. Any remaining stains may be acidic or comprising environmental dirt and will therefore require washing and/or the use of a chelating agent to break down salts. When choosing solvents, paper conservators not only need to consider the composition of the stain (i.e. tape adhesive) but also the paper, particularly if it is modern and includes materials such as brighteners and plasticisers. Although at times making a solvent mixture is suitable, certain properties inherent to mixtures can be problematic for some works and a direct single greener solvent replacement may be more desirable.

Many solvents (found in pertinent literature) that help to remove tape stains are of a low to middle polarity and are also quite harmful including toluene, benzene, tetrahydrofuran, N,N-dimethylformamidemethylformamide and xylene (Smith *et al.* 1984; Cannon 2011).

Choosing a new greener alternative to replace a toxic solvent is complex and takes time to research therefore it can be helpful to look in other industries and applications. Studying relevant research from the scientific community was my first step in finding an alternative solvent with comparable HSP to toluene. The journal, *Green Chemistry* by the Royal Society of Chemistry has up-to-date research on the subject and is a very useful resource. Similar societies exist worldwide such as Gesellschaft Deutscher Chemiker (GDCh), the Korean Chemical Society (KCS), and the Royal Australian Chemical Institute (RACI). It is worth contacting your chemical supplier: Merck, for example, has a very useful tool to evaluate sustainable chemicals using the 12 principles of green chemistry.

You may also find a solvent using the Chem21 guide. The updated version of the GSK (Glaxo Smith Kline) Solvent Sustainability guide has many neoteric solvents and attempts a life-cycle analysis on most of the solvents, making it is easier to assess the whole environmental picture rather than specific individual qualities. An article published in *Green Chemistry* in 2016 is also very useful (Alder *et al*. 2016).

As illustrated above, when considering one or several solvents, they can be plotted on the Teas chart in order to gain an idea of where it sits. Surrounding solvents may also be useful to evaluate if the neoteric solvent is better than existing solvents in the solvent cabinet as a comparison. It is crucial that testing is carried out on samples, using similar materials and age-testing, and test as much as possible. Even if equipment is limited, this does not prevent you from testing – the process may just take longer. The samples should be similar to the treatment process to be undertaken. If no oven is available, the samples can be aged in bright light and heat. The result should be examined under magnification and in different light sources, i.e. UV and raking light. Share and compare results with others. Strength-testing paper is vital, particularly when working with modern papers. Investigate the artwork's technical materials and try to match these in the mock-up when testing. Various analytical techniques can also be undertaken to establish whether a new solvent is suitable for conservation, and combined with information from other sources, such as Hansen solubility parameters.

Research is always ongoing and, in my case, speaking to academics and researchers has developed into some fruitful relationships and produced new answers.

Melissa Lewis

For both approaches (i.e. Option 1 or 2), the initial consideration of some other factors is also beneficial in practice. For instance, substrate concentration has an effect on solubility, and viscosity also varies depending on where in the polymer solubility window the solvent (or mixture) is located. The position of a solvent (or mixture) in the solubility window of a polymer and solvent evaporation rates have a marked effect on the dried film characteristics of the polymer. As mentioned above, the RER is perhaps the key physical property to be assessed, so, where these data are available, they are included for the solvents in Table 1 (pp. 42–3) and Table 2 (pp. 44–5).

STEP 3 RESEARCH AND TESTING

'However much data we supply, and however good the automatic estimation tools might be, there is always a need for measurements of real-world materials' (Hansen 2021). Whether Option 1 or 2 is applied in Step 2, research and careful testing of the proposed replacement solvent/mixture prior to treatment is critical. Alongside the testing to be carried out by the conservator is a need for in-depth research by cultural heritage scientists into potential long-term, 'invisible' solvent effects. In considering the potential introduction of novel mixtures and neoteric solvents, and for certain substrates in particular, the acquisition of equivalent data and information is an important requirement.

Accordingly, potential resin-based recipes have not been provided, partly because what may be an appropriate solution for a paper substrate may be problematic for another substrate. For instance, cyclic aprotic solvents (such as TMO) interact well with free fatty acids and are therefore generally considered inadvisable substitutes for treating paint films. The specific sensitivity of any original object must therefore include the consideration of relevant research data and be verified by the conservator *in situ*. An understanding of the original material response in the specific artwork has to be determined if conservation materials are to be applied with safe solutions. By using a standardised testing methodology, comparative empirical data can be generated, thereby increasing the possibility for determining safer, greener recipes for specific types of substrate (Garcia 2014; Fife 2020).

These three steps are cyclic – having eliminated one hazardous solvent you can look at the next. While designed for immediate application, the conscious methodology behind the approach outlined in these three steps is to enable the continued incorporation of the latest research and information. It is hoped that such work and its subsequent dissemination – not least through SiC's 'Greener Solvent' project – will help ensure the correct application of appropriate replacement solvents and mixtures in conservation.

Table 1 Selected ratings and properties of certain classical (traditional) organic solvents in 'Example Solvent Cupboard'. Many of the traditional solvents already in use (e.g. simple alcohols, esters) come under a recommended listing, clarifying that *'green' does not mean new*. Data sources for Hansen parameters: Horie 2010; Abbott 2013. Physical properties data sources: Horie 2010; pubchem.ncbi.nlm.nih.gov; Rivers and Umney 2003. Fractional parameters provided in the above sources or calculated by G.R. Fife based on Hansen parameters.

Solvent	Selected solubility parameters		CHEM21 rating	Relative evaporation rate to n-butyl acetate nBuAc = 1 (associated definition)*	Vapour pressure kPa@^{0}C	Boiling Point ^{0}C	STRUCTURE
Horie#/ Name (assigned letter if not referenced therein)	Hansen Solubility $\delta_d \delta_p \delta_h$ MPa½	Fractional $100f_d$ $100f_p$ $100f_d$	Safety/Health/ Environment S,H,E #'s				
1. Iso-octane 2,2,4-trimethylpentane	15 0 0	100 0 0	**predicted problematic	c. 2.8 (medium)	5.41@21	99	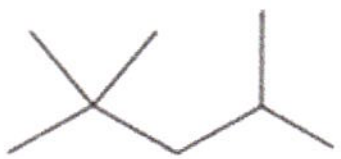
7/4. Toluene (methylbenzene)	18 1.4 2	84 7 9	problematic 5,6,3	2.3 (medium)	2.93@20	110	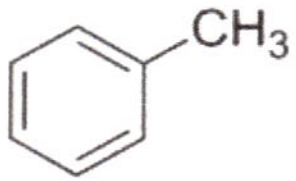 CH_3
8,9/6. Xylene (dimethylbenzene isomers)	17.6 1 3.1 (p-isomer)	81 5 14	problematic 4,2,5	c. 0.75 (slow)	1.07@20	138	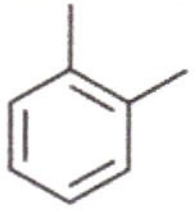 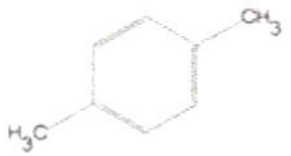 CH_3 H_3C isomers o-& p- shown
47/14. Ethylacetate	15.8 5.3 7.2	56 19 25	recommended 5,3,3	4.3 (fast)	10.13@20	77	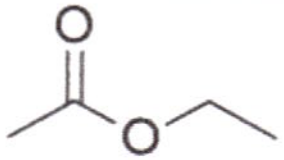 O O
49/16. n-butylacetate	15.8 3.7 6.3	61 14 24	recommended 4,2,3	1 (medium)	1.2@20	126	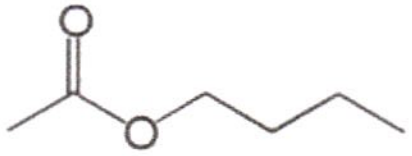 O O

63/28. Acetone (propanone)	15.5 10.4 7	47 32 21	recommended 5,3,5	7.8 (fast)	24.13@20	56	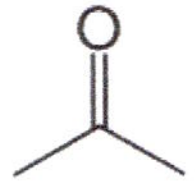 O
64/26. Methyl ethyl ketone (butanone isomer)	16 9 5.1	53 30 17	recommended 5,3,3	4.6 (fast)	10.5@20	80	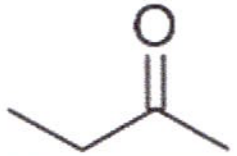 O
67/18. Methyl isobutyl ketone (hexone isomer)	15.3 6.1 4.1	60 24 16	recommended 4,2,3	1.4 (medium)	2.13@20	117	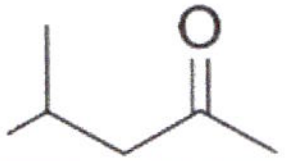 O
75/44. Ethanol 99.8%	15.8 8.8 19.4	36 20 44	recommended 4,3,3	2.4 (medium)	5.8@20	78	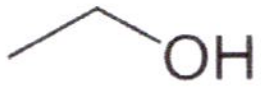 OH
77/39 2-Propanol (isopropanol)	15.8 6.1 16.4	41 15 43	recommended 4,3,3	1.7 (medium)	4.4@20	82	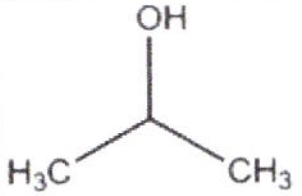 OH H_3C CH_3
78/32. 1-butanol 99+% (butanol isomer)	16 5.7 15.8	43 15 42	recommended 3,4,3	0.46 (slow)	0.53@20	118	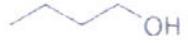 OH
Mixture D. Shellsol D40 (paraffinic and naphthenic hydrocarbons C9-C11.)	?	96 2 2	**predicted problematic	0.18 (slow)		162–192	

Notes

*General evaporation-rate-definitions relative to n-butyl acetate (where nBuAc = 1): fast > 3 , medium > 0.8 < 3, slow < 0.8. In addition to this, or alternatively, an estimate for 'seconds to 90% evaporation' can be helpful for envisaging the evaporation rate (Rivers and Umney 2003). This can be approximixated by dividing 458 by the nBuAc# (where nBuAc = 1), e.g. For a solvent with nBuAc # 7.8: 458/7.8 = c.59 seconds to 90% evaporation. On this basis the above definition becomes: fast < 153 seconds, medium between 153 and 573 seconds, slow > 573 seconds.

**Predicted – not included in CHEM21 guide but conclusion generally assigned in accordance (default problematic). Absolute evaporation rate (e.g. mass/time units) is dependent on temperature, atmospheric pressure, humidity, air flow, viscosity etc. *Relative evaporation rate is a guide only.* Typically the higher the vapour pressure, the lower the boiling point and the faster the evaporation rate.

Table 2 Selected ratings and properties of neoteric (modern) organic solvents considered for the 'Example Solvent Cupboard'. While the solvents noted here have undergone some initial examination for potential introduction into conservation practice, their inclusion in this list is not a recommendation. *Furthermore, solvents with a CHEM21 health rating 5 or above were immediately excluded from inclusion in Figure 2. Flash point data were not included here but high flammability (extremely low flash points) should be noted in SDS.* Fractional parameters provided in Horie or calculated by G.R. Fife based on Hansen parameters. Data sources for Hansen parameters: Horie 2010; Abbott 2013, 2021; Sherwood *et al.* 2014; Byrne *et al.* 2017; Jalan *et al.* 2019. Physical properties data sources: Horie 2010; pubchem.ncbi.nlm.nih.gov; ThermoFischer, Sigma-aldrich, Gaylord Safety Data Sheets.

Solvent Horie#/ Name (assigned letter if not referenced therein)	Selected solubility parameters		CHEM21 rating Safety/Health/ Environment S,H,E #'s	Relative evaporation rate to n-butyl acetate nBuAc = 1 (associated definition)*	Vapour pressure kPa@⁰C	Boiling Point ⁰C	STRUCTURE
	Hansen Solubility $\delta_d\ \delta_p\ \delta_h$ MPa½	Fractional $100f_d$ $100f_p$ $100f_d$					
C. Cyclopentylmethylether (CPME)	16.7 4.3 4.3	66 17 17	problematic 7,2,5	No data available	4.27@25	106 (initial)	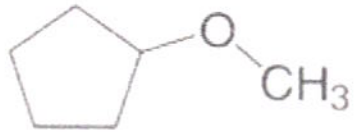
M. Methyl-tetrahydrofuran (MeTHF)	16.9 5 4.3	65 19 16	problematic 6,5,3	No data available	13.6@20	78 (initial)	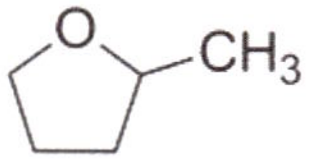
T.TMO (2,2,5,5-tetramethyl-tetrahydrofuran)	15.4 2.4 2.1	77 12 11	in testing (AD)	No data available	No data available	112	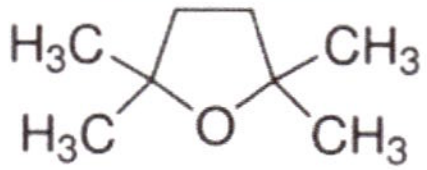
L. d-Limonene ((4R)-1-methyl-4-prop-1-en-2-ylcyclohexene)	17.2 1.8 4.3	74 8 18	problematic 4,2,7	No data available	0.21@20	176 (initial)	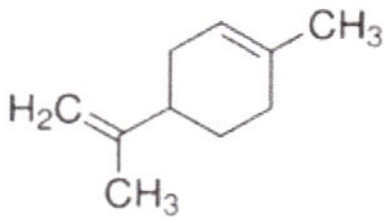
Y. Cyrene ((1S,5R)-6,8-dioxabicyclo[3.2.1] octan-4-one)	18.8 10.6 6.9	52 29 19	problematic 1,2,7	No data available	0.028@25	227	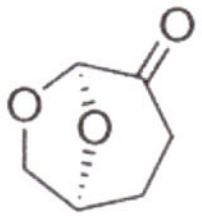

44. 1-methoxy-2-propanol	15.6 6.3 11.6	47 19 35	**predicted problematic	0.71	1.2@20	120	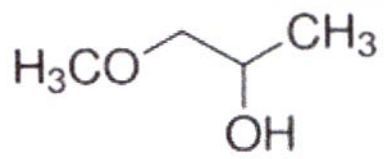
54. Ethyl lactate (ethyl 2-hydro-xypropanoate)	16 7.6 12.5	44 21 35	problematic 3,4,5	0.22	0.23@20	154	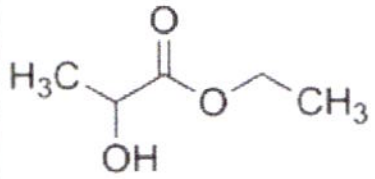
123. Dimethylsulphoxide (DMSO)	18.4 16.4 10.2	41 36 23	problematic 1,1,5	0.026	0.06@20	189	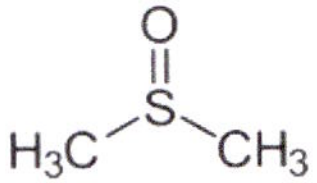
58. Dimethylcarbonate	15.5 3.9 9.7	53 13 33	recommended 4,1,3	3.22	5.6@20	90	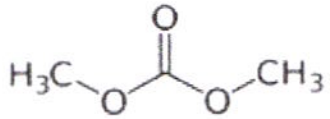
PC. Propylene carbonate	20 18.0 4.1	47 43 10	problematic 1,2,7	No data available	0.005@20	242	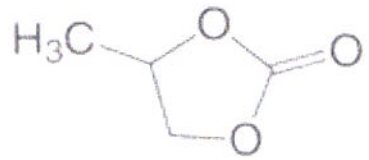
P. p-Cymene (4-isopropyltoluene)	17.3 2.4 2.4	78 11 11	problematic 4,5,5	No data available	0.2@20	177	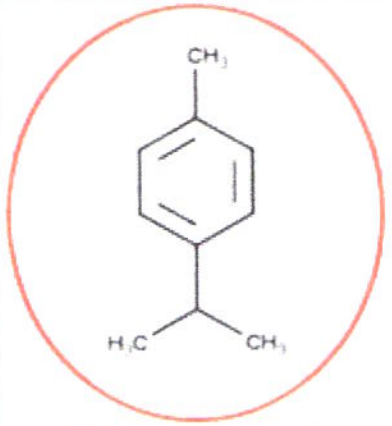
γ-Valeractone ((5R)-5-methyloxolan-2-one)	No data available		problematic 1,5,7	No data available	No data available	No data available	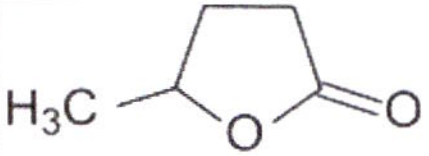

Notes

*General evaporation-rate-definitions relative to n-butyl acetate (where nBuAc = 1): fast > 3 , medium > 0.8 < 3, slow < 0.8. In addition to this, or alternatively, an estimate for 'seconds to 90% evaporation' can be helpful for envisaging the evaporation rate (Rivers and Umney 2003). This can be appromixated by dividing 458 by the nBuAc# (where nBuAc = 1). For example, for a solvent with nBuAc # 0.71: 458/0.71= c.645 seconds to 90% evaporation.

**Predicted – not included in CHEM21 guide but conclusion generally assigned in accordance (default problematic) unless otherwise advised (AD). Absolute evaporation rate (e.g. mass/time units) is dependent on temperature, atmospheric pressure, humidity, air flow, viscosity etc. Relative evaporation rate is a guide only – typically the higher the vapour pressure, the lower the boiling point and the faster the evaporation rate.

Table 3 Determining the fractional parameters of two example mixtures EI and ED – as illustrated in Figure 3 – that come close to the target region for replacing toluene: fd = 84, fp = 7, fh = 9. EI is a 2:1 mixture of isooctane:ethylacetate, ED is a 4:1 mixture of Shellsol D40:ethanol. Solvents numbered according to Horie 2010 and letter assignment as in Tables 1 and 2. Their inclusion here is to demonstrate methodology, not as a recommendation of any mixture or for any specific application.

Solvent Horie# Name (assigned letter if not referenced therein)	Fractional Parameters x (proportion in mix)		
	100fd	100fp	100fh
47. Ethylacetate	56 x (1/3)	19 x (1/3)	25 x (1/3)
1. Isooctane	100 x (2/3)	0 x (2/3)	0 x (2/3)
MIX EI 2:1 Isooctane:ethylacetate	c. 85	c. 6.5	c.8.5
75. Ethanol	36 x (0.2)	20 x (0.2)	44 x (0.2)
D. Shellsol D40	96 x (0.8)	2 x (0.8)	2 x (0.8)
MIX ED 4:1 Shellsol D40:ethanol	c. 84	c. 6	c. 10

Figure 1 Teas chart with classical (traditional) organic solvents in the 'Example Solvent Cupboard' from Table 1. Solvents numbered according to Horie 2010 and letter assignment as in Table 1. Inclusion here is not a recommendation.

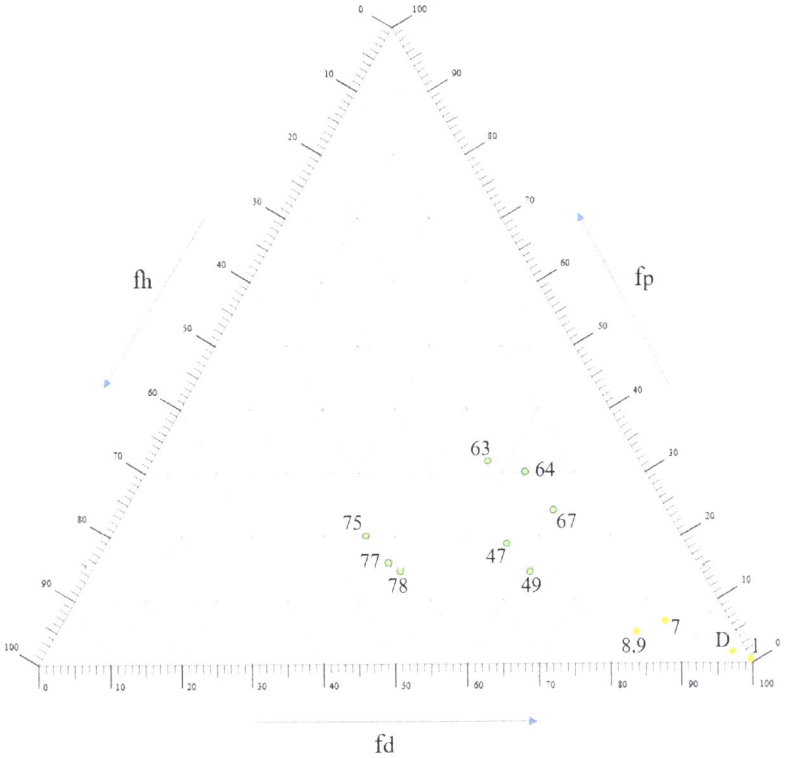

Notes

Problematic: 1 = iso-octane (2,2,4-trimethylpentane), D = Shellsol D40 (paraffinic and naphthenic hydrocarbons C9-C11), 7 = toluene (methylbenzene), 8/9 = xylene (dimethylbenzene isomers).

Recommended: 47 = ethylacetate, 49 = n-butylacetate, 63 = acetone (propanone), 64 = methyl ethyl ketone, 67 = methyl isobutyl ketone (hexone isomer), 75 = ethanol 99.8%, 77 = 2-propanol (isopropanol), 78 = 1-butanol 99+%.

Figure 2 Teas chart with combined classical (traditional) and neoteric (modern) organic solvents from Tables 1 and 2. Where previously undefined, published HSP were used to generate fractional parameters and assign Teas chart positions here. Solvents numbered according to Horie 2010 and letter assignment as in Tables 1 and 2. Inclusion here is not a recommendation and solvents with a CHEM21 health rating of 5 or above were immediately excluded.

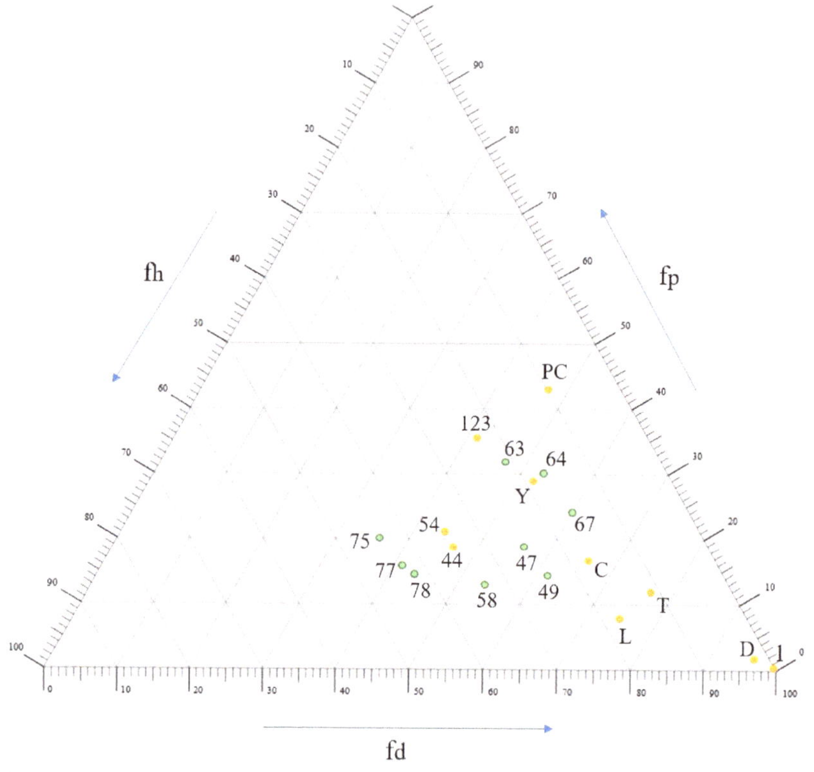

Notes

Problematic: **Problematic**: 1 = iso-octane (2,2,4-trimethylpentane), D = Shellsol D40 (paraffinic and naphthenic hydrocarbons C9-C11), T = TMO (2,2,5,5-tetramethyltetrahydrofuran), L. d-limonene ((4R)-1-methyl-4-prop-1-en-2-ylcyclohexene), C = cyclopentylmethylether (CPME), Y = cyrene ((1S,5R)-6,8-dioxabicyclo[3.2.1]octan-4-one), 44 = 1-methoxy-2-propanol, 54 = ethyl lactate (ethyl 2-hydroxypropanoate), 123 = dimethylsulphoxide (DMSO), PC = propylene carbonate.

Recommended: 47 = ethylacetate, 49 = n-butylacetate, 58 = dimethylcarbonate, 63 = acetone (propanone), 64 = methyl ethyl ketone, 67 = methyl isobutyl ketone (hexone isomer), 75 = ethanol 99.8%, 77 = 2-propanol (isopropanol), 78 = 1-butanol 99+%.

Figure 3 Illustration of an alternative mixture determination on the Teas chart. Lines between CHEM21-recommended classical (traditional) and remaining solvents from Table 1 indicate the combinations from the 'Example Solvent Cupboard' that could come close to the target area for replacing toluene: fd = 84, fp = 7, fh = 9. Solvents numbered according to Horie 2010 and letter assignment as in Tables 1 and 2.

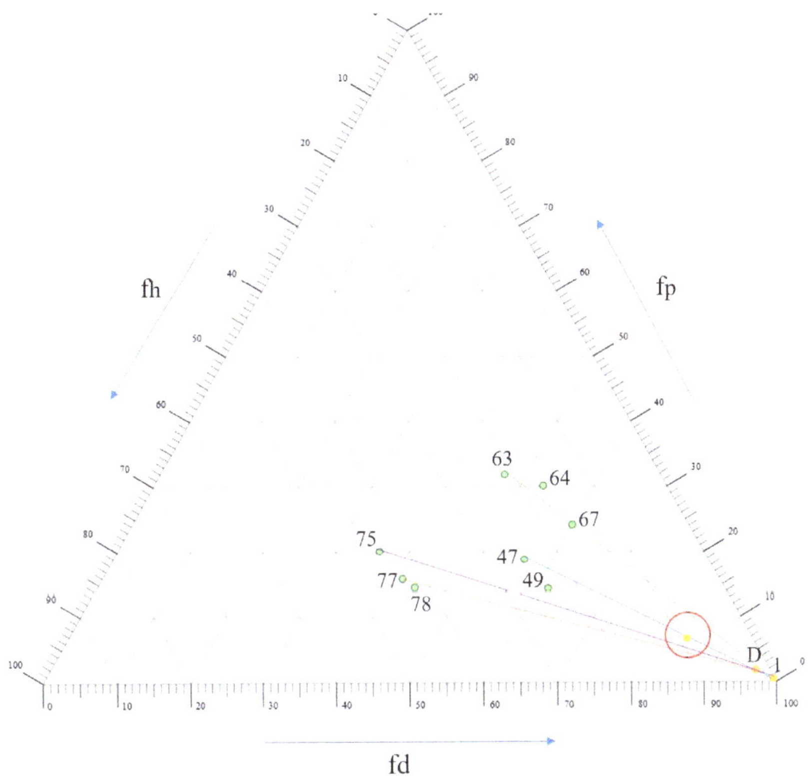

Notes

Problematic: 1 = iso-octane (2,2,4-trimethylpentane), D = Shellsol D40 (paraffinic and naphthenic hydrocarbons C9-C11).

Recommended: 47 = ethylacetate, 49 = n-butylacetate, 63 = acetone (propanone), 64 = methyl ethyl ketone, 67 = methyl isobutyl ketone (hexone isomer), 75 = ethanol 99.8%, 77 = 2-propanol (isopropanol), 78 = 1-butanol 99+%.

Explanatory notes

Although the following information is not critical for carrying out these greener steps, they are points related to accuracy and further interest.

STEP 1 STUDIO SOLVENT CUPBOARD ASSESSMENT

Greener solvent selection

- For greener solvent selection CHEM21 is introduced here as the most straightforward step to apply, but other approaches are available (such as the online tool from the American Chemical Society, and information from regulators such as REACH).
- Developed by a European consortium, CHEM21 was based on a survey of publicly available solvent selection guides worldwide. 'In order to rank less classical solvents, a set of Safety, Health and Environment criteria is proposed, aligned with the Global Harmonized System (GHS) and European regulations. A methodology based on a simple combination of these criteria gives an overall preliminary ranking of any solvent. This enables in particular, a simplified greenness evaluation of bio-derived solvents' (Prat *et al.* 2016: 288).
- *For any assessment of greenness, the specific application (manner of use) is critical* and is one of the reasons why generating full LCAs on a single solvent is highly complex. Therefore for our purposes, a selection tool such as CHEM21 is not perfect, with the biggest issues relating to assessing safety and irrelevance of distillation recovery: 'for some of these applications in which the solvent is not recovered, the boiling point impact can be revised. In the same way, the flash point impact needs to be scored more severely for applications using solvents in open air such as paint stripping, coating, etc. This will reflect the high interest of some bio-derived solvents for such applications, whereas these solvents often appear as "problematic" for pharmaceutical chemistry, as a result of their high boiling points which complicate the recovery and downstream processing on scale, or require the use of new process technologies' (Prat *et al.* 2016: 294).
- It should be reassuring that many of the traditional solvents being used already (e.g. simple alcohols, esters) come under a '**Recommended**'

listing, and that for the field at large it also clarifies that *green does not mean new.*

- Rather it is clear that serious contenders are needed for replacing low polar petrochemical-industry based, hydrocarbon, and distillation-derived mixtures. For now, retaining some of these in our cupboard is likely to be necessary for the creation of the mixtures needed for certain applications; not the perfect solution, but an improvement.
- Depending on availability, different such petrochemical-derived mixtures will be variously preferable for use. Nomenclature varies (Phenix 2007) but generally properties are linked to boiling point ranges and the potential for low aromatic content.

STEP 2 WHAT NOW? SEEKING SUBSTITUTES AND ALTERNATIVES

Modelling considerations

- Since everything we have ever learned about how the world works is based on a model we have been told about or developed ourselves from experience, some further discussion is provided here.
- Solvent modelling continues to evolve with arguably the most refined method currently available being COSMO-RS, which enables the chemical potential differences of molecules in liquids (the chemical ENERGY) to be calculated and transformed into properties such as solubilities, activities or vapour pressures.
- As previously mentioned, when considering solvents in conservation there is a tendency to focus on cleaning applications and their possible effects on the original substrates, in which many potential solvent actions can be involved. With the individual nature and general complexity of our practical situations (as evidenced by research over the last 70 years on paint films) in-depth discussions and all potential modelling approaches are outside the scope of this handbook.
- However, in discussing coating applications (e.g. the application of an individual polymer we want to use as a consolidant or say, a varnish) we are in the theoretical modelling realm of dilute solutions. Within this context the Hansen model is sufficiently accurate and even the empirical Teas chart (or alternative presentations in Triansol, Tri Solv, Solvent Solver) – based on HSP – are reliably applicable as outlined below (Hansen 2021).

- HSP are thus specifically considered of importance within this chapter for our focus on developing greener alternative solutions in our recipes. In 1966, Charles Hansen developed the theory for his parameters which divide the total Hildebrand value into three component forces (dispersion, polar and hydrogen bonding). Hansen calculated the dispersion force for a particular liquid using what is called the homomorph method and separated the polar value into polar and hydrogen bonding component parameters.
- The coordinates at the centre of a solubility sphere are located by means of the three component parameters (d,p,h), and the radius of the sphere – called the interaction radius (R) – is indicated. The distance between two molecules, conventionally called Ra, is the measure of how alike they are: the smaller Ra, the more likely they are to be compatible. From measurements of 'good' and 'bad' solvents, a sphere of radius Ro for the solute can be defined. Solubility is indicated when the relative energy difference (RED), i.e. Ra/Ro, is less than 1.
- This model remains one of the most widely accepted to date and the parameters are accurate for dilute solutions, to the point that for *bio-based solvent certification within the EU there is a requirement to provide the HSP of the solvent.* For mixtures, the HSP of a solvent mixture depends on the relative amounts of the individual solvents present, and becomes predictably 'good' for a solute even if the individual solvents are not. Creating synergistic mixtures – as proposed here – has similarly been used to replace undesired solvents, to reduce costs and improve performance at the same time, within other industries.
- The HSP for individual solvents are generally (although not always) available, and there is an associated HSPiP tool which provides the potential for a fully updated/revised set of HSP data for the 1200+ chemicals that constitute the original standard Hansen set with HSP and other data tabulated for another 10,000 chemicals. However, this latter lies outwith most conservation resources (Hansen 2021).
- As mentioned above, when creating solvent mixtures the HSP of the mix can be calculated from the volume-weighted average of the HSP of the components (Abbott 2021). Since the manual calculations are unwieldy, a useful resource for this is the Hansen Space provided by Chris Stavroudis in his Modular Cleaning Programme (MCP) (Stavroudis and

Doherty 2013). This incorporates consideration of molar volumes and density of the individual solvents within created mixtures, enabling a prediction of solubility with certain specific solutes included within the programme. The Teas fractional parameters of the mixtures are also provided based on two calculations: the relative combination of the individual solvents on the Teas chart directly (i.e. with a line between the two solvents), and those derived from the HSP of the mixture. As anticipated, the greater the difference in molar mass and density of the individual solvents within the mixture, the more obvious the discrepancy between the resulting Teas chart locations.

- This brings us to an important factor with solvent mixtures, namely non-ideal behaviour. Since the initial identification of a peak oil swelling region and its placement on the Teas chart, research has repeatedly confirmed that a binary mixture of solvents may have a greater swelling effect on the original paint than either of the two individual solvents used alone (e.g. Hedley 1980; Phenix 2002, 2013; Zumbühl 2012). The larger the difference in polarity between two mixed solvents, the greater is this observed deviation from ideal behaviour, and this deviation correlates with the boiling points of the mixtures. A fuller explanation related to a lowering of the cohesive energy within the mixture is well described elsewhere (Zumbühl 2019). When preparing mixtures this effect should be considered in the initial solvent selection.

Empirical tools for solutions in practice

- In a continuing line of decreasing resistance to resources comes the Teas chart. Please note that this is not a negative connotation – in fact from a practical conservation viewpoint it is quite the opposite. Conspicuously aware of all its theoretical failings – its inadequacy in modelling solvent effects on paint films in particular have been well described (Phenix 2002, 2013; Baij *et al.* 2020) – the Teas chart has remained a remarkably useful engineering tool and thereby popular within the field. The lack of theoretical foundation has not prevented it from being useful in preparing dilute solutions and for this it remains a convenient method for conservators.
- Teas developed the familiar triangular graph in 1968 presenting the relative amounts of the three component forces. Solvent positions were

originally located on the graph according to Hansen values and what the approach theoretically lacked was initally supplemented by empirical testing. The placement of the newer solvents on the Teas chart presented here are based solely on published Hansen parameters.

- In summary, we can consider the Hansen parameters when selecting a specific solvent to dissolve a conservation material, and we have the power to manipulate our solvent recipe on the TEAS chart. Examining the solubilities of polymers, alongside consideration of solvent characteristics and the sensitivity of the original materials, can thus enable tailoring of recipes by conservators themselves.
- For specific solvent selection in this regard, the book by Stefan Zumbühl (2019) also contains plots of associated research data. The dissolution rate of conservation varnishes is provided with generally strong acceleration at vapour pressures above 100hPa (25 °C). Although the data cannot aid with determining mixtures, they can still be useful for considering varnish applications when selecting the most effective solvent for a specific material and avoiding those for which the original materials are known to be sensitive.

A case study in analytical approaches for solutions in practice

One route for specific solvent selection with greater resources can be through analytical studies, for instance as occurred after Mark Rothko's painting *Black on Maroon* from 1958 was damaged in 2012 by a vandal using black graffiti ink. Through research at Dow in collaboration with a conservation scientist from the Tate, the HSP of the ink were determined and solubility predictions in the form of relative energy density (RED) values were generated. From over 600 solvents, 16 with high ink solubility predictions (i.e. low RED values) were identified, 5 of which were included within a larger group for subsequent empirical testing and solvent refinement (Barker and Ormsby 2015). Beyond this associated example, the methodologies, applied in collaborations between Dow, Tate and GCI, for the rapid discrimination and screening of the cleaning efficacy of possible cleaning liquids and the evaluation of the effects of these liquids on representative test samples of artists' materials, further illustrate the potential benefits of such studies for the field at large (Keefe *at al.* 2011; Ormsby *et al.* 2016; Wills *et al.* 2021).

Neoteric solvent introduction

- As mentioned above, many of the neoteric 'green' solvents introduced recently in conservation have been termed so due to their production from biomass. They may pose potential solutions for particular industrial reactions, but their production is not always based on what *has* to be replaced, and not with regards to our conservation demands. When examining parameters it is clear that in many instances they may be unnecessary additions to a conservator's solvent cupboard. Continuing our perspective at the start of the chapter (the removal of toluene and xylene) we can see that the HSP of TMO comes close. D-limonene is also in the general area of interest, and p-cymene is also close (but CHEM21 health rating of 5 negates its inclusion in Fig. 2, p. 48, here).
- With either a single greener replacement or greener alternative mixture approach we aim to save the world *and* the art – and last, but certainly not least, without harming ourselves. This means that in addition to good scores for low energy and renewable sources in their manufacture, toxicology data are critical, especially for introducing the new solvents that we need for our field. It should be noted that being named 'bio' or 'green' does not guarantee low toxicity.
- In industrial processes, it might be smart to introduce a biorenewably sourced solvent as a slightly less harmful direct replacement. While robust health and safety procedures should be paramount, conservators may still have a working exposure whereby a health rating of 5 or above would negate introduction to our hypothetical solvent cupboard.
- One reason certain solvents such as ethyl lactate are defined as 'problematic' in the CHEM21 review is due to their high boiling point (which affects their rating considering recycling via distillation recovery methods) (Prat *et al.* 2016). These are aspects that could be adjusted by further research and assessment with regard to our field in particular. 'Armed with the thorough comparison of potentially greener solvents, the user can make an intelligent choice, including factors such as price, to arrive

at an overall best-practice decision. Is this the perfect way to save the planet? No. Is it the smart way? Yes' (Abbott 2021: 76).

- While we have to rely on others for the provision of toxicology data, the assessment of potential effects on the artwork is the domain of the conservator and cultural heritage scientist. For new or greener-labelled solvents we should question what the potential effects are, both visible and long term, on the artworks and artefacts in our care. Since it is our job to consider this, how can we improve our risk assessment and risk reduction?
- This brings us to the importance of testing both the individual artwork and substrate type. For many of the new or greener-labelled solvents potentially for use within conservation there has been little rigour investigating their short and longer term effects, despite this *critical information being vital to a balanced risk assessment.*

STEP 3 RESEARCH AND TESTING

Importance of research and testing

- Although in the practical conservation world at large, analytical aid for examining a specific object is often not available, heritage science research and the associated data have been highly beneficial in informing our selection processes and practices. For instance, the information provided over the last 70 years or so by investigations into the effects of traditional solvents on paint films has proved invaluable.
- While it is likely that we will always be limited in the application of solubility models and related research data directly to our conservation practice (since specific interactions and behaviour, even across an apparently homogeneous surface, will vary), their use in conservation and scientific research, and our consideration of these developments, can lead to profound insights.
- With the individuality of each situation there also remains a need for the very small tests we are obliged to carry out on the specific work or samples prior to embarking on solvent-related treatments. These critical observations are key to how we understand an artwork and select the 'appropriate' solvents and mixtures, and why our own testing is so important. The conservator's active empirical tests provide data for their conceptual modelling, helping to reduce the risks of

solvent use in both removing and applying coatings. They develop our 'heuristic' model of the behaviour of the individual artwork, with the added advantage that, with every new situation, we add to our dataset, thereby enhancing our conceptualized modelling of paint film behaviour in general (we could equally term this our 'experience').

- In terms of selecting solvents safe for a specific individual work, it is most certainly true that these data collected during treatment are critical. A fundamental consideration is to determine the relative rate of dissolution of the material to be dissolved versus the potential actions on the original materials in the original artwork being treated. *Safely increased efficiency of solvent use whereby less is used is inherently 'greener'.*
- A relatively simple system for measuring the solubility of unidentified non-original materials on the surface of the art object has been suggested by extracting microsamples of the material, examining them under the microscope, and then determining the amount of time it takes a sample to swell or not in different solvents (Zumbühl 2019).
- Approaches such as measuring solvent quantities and timing contact, as in the Solvent Star (Fife 2020), can also clarify solvent efficiency and inform on safety, giving added confidence in both observations and final approach selection.

Further considerations and research questions

The development, research and testing of new solvents for those that we need to replace within conservation (e.g. petrochemical-derived mixtures, aromatic functionality with problematic CHEM21 rating) is vital. Valid assessment of such (new) green(er) replacement solvents requires both analytical and empirical approaches: ideally combining in-depth research from cultural heritage science, the solubility data of specific conservation materials, and the comparative data collected by conservators from their individual works. It is also hoped that the ultimate introduction of replacement novel solvents (when required) will be facilitated by the open access dissemination possible within the planned Phase 2 of the 'Sustainability in Conservation Greener Solvents' project. Accurate and relevant research regarding these potential solvents can be provided thereby ensuring their correct application into conservation practice.

Continuing work regarding LCA considerations for the diverse and specific applications of solvents within conservation will be helpful in determining their relative impact and refining solvent rankings in our field. A relatively simple but critical aspect regarding greener solvent use is correct waste disposal – for either incineration or recycling this entails the separation of aqueous and organic solvent wastes (with further separation into non-halogenated/halogenated solvents if these are used) in the studio.

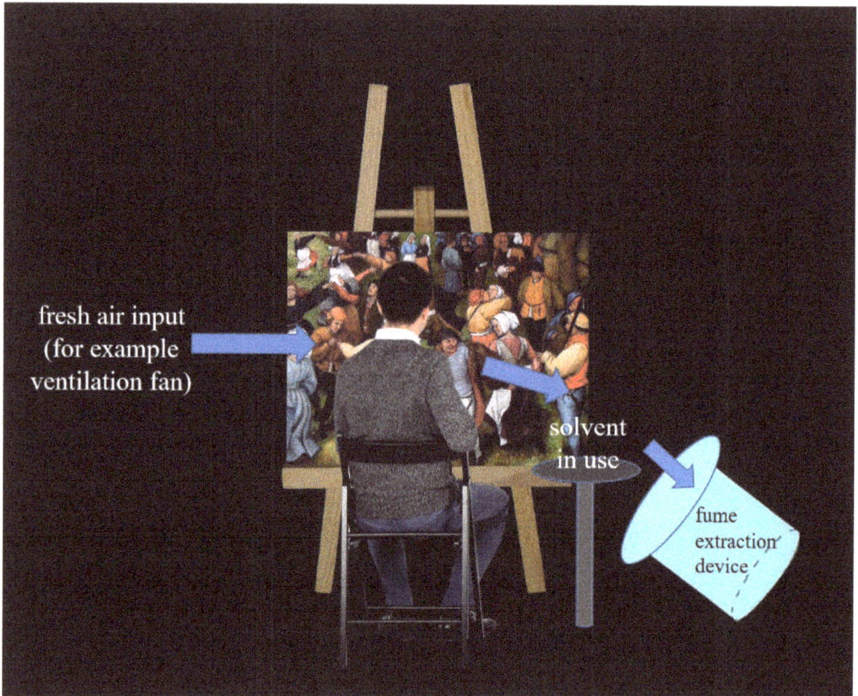

Figure 4 Potential work set-up. Organic solvents have a higher vapour density than air – an important factor for your work set-up and cleaning any accidental spillages. Consider the air flow direction, increased ventilation and extraction position below the solvent location.

Checklist

- Do I need to use a solvent?
- Can I use less solvent? (For instance by using the solvent most safely efficient for the application and/or adapting the application technique.)
- Is the solvent '**Recommended**' on CHEM 21? What is the H number?
- Check with the supplier that the product is actually biomass-based.
- Check the requirements for safe use and the health and safety classifications.
- Have I correctly considered my work set-up?
- What are the considerations for safe storage, correct disposal/recycling?

Conclusions

Research in conservation has developed our understanding of solvent behaviour and effects, helped the fine-tuning of our organic solvent-based treatment approaches (varnish removals/ applications/ consolidations), and highlighted the importance of reduced quantities and contact. This last is a fundamental truth equally applicable to us and the planet. Health is critical and where we need to use organic solvents (or for scenarios where more traditional approaches pose little risk) *we can aim to use less, and in a manner as safely efficient as possible.* Avoiding solvent use where possible and eradicating the most hazardous solvents from use through greener alternative solutions should be primary aims and immediately achievable. Despite the critical importance of the climate crisis, proposing disruptive events in our work systems are unlikely to prove a successful approach. Rather it is frequently through the culmination of steady consequential advances, however small, that progress is made, especially if these steps are taken by many. On this basis, it is hoped that you will feel encouraged to dare to open your solvent cupboard and take that first 'greener step' today.

Glossary

Aqueous solvents Solutions where water is the solvent are called aqueous solutions.

Aromatic Compounds that contain one or more rings with delocalized pi electrons in resonance all the way around them.

Atom economy (atom efficiency/percentage) The conversion efficiency of a chemical process in terms of all the atoms involved and the desired products produced.

Binary mixture A binary mixture consists of two types of molecules.

Bioaccumulation A process of accumulation of chemicals in an organism that takes place if the rate of intake exceeds the rate of excretion.

Bio-based material A material intentionally made from substances derived from living (or once-living) organisms.

Biorenewable Able to be renewed via biological means; produced by biological organisms.

Chemical potential differences The difference in chemical potential between two locations or a chemical potential gradient. It is the driving force for the migration of the corresponding chemical species from high chemical potential regions to lower chemical potential regions.

COSMO-RS The COnductor-like Screening MOdel for Realistic Solvents calculates thermodynamic properties of fluids and solutions based on quantum mechanical data.

Degradation In chemistry it is the act or process of simplifying or breaking down a molecule into smaller parts, either naturally or artificially.

EHS Environmental health and safety.

Empirical testing A research method that employs direct and indirect observation and experience.

Globally Harmonized System (GHS) The United Nations uses this as a form of classification and labelling of chemicals.

Green solvents Green solvents are traditionally defined as environmentally friendly solvents, or biosolvents, which are derived from the processing of agricultural crops.

Greener solvent alternative approach Attempts to eliminate undesirable solvents by seeking structurally related compounds not yet covered/ restricted by the legislative and regulatory measures.

Hansen solubility parameters (HSP) Hansen solubility parameters were developed by Charles M. Hansen in his PhD thesis in 1967 as a way of predicting if one material will dissolve in another and form a solution.

Hydrocarbons Hydrocarbons are organic compounds that consist of hydrogen and carbon; they can be either saturated or unsaturated.

Inorganic Relating to or denoting compounds which are not organic (broadly, compounds not containing carbon).

Non-ideal behaviour (in solutions) Ideal solutions are where the interactions between the molecules are identical (or very close) to the interactions between molecules of different components; the forces between the particles in the mixture are exactly the same as those in the pure liquids. This is not the case in non-ideal mixtures which result in various behaviours and effects.

Organic solvents Organic solvents are carbon-based volatile compounds.

Petrochemicals The chemical products obtained from petroleum by refining. Note that some chemical compounds made from petroleum can also be obtained from other fossil fuels, such as coal or natural gas, or renewable sources such as maize, palm fruit or sugar cane.

Polarity In chemistry, polarity is a separation of electric charge leading to a molecule or its chemical groups having an electric dipole moment, with a negatively charged end and a positively charged end. While molecules can be described as 'polar covalent', 'non-polar covalent', or 'ionic', these terms are often relative, with one molecule simply being more polar or more nonpolar than another. The distribution of electric charges in a molecule can also be affected by its surrounding environment.

Polymer solubility window The graphically represented boundary of solubility according to the locations of solvents and non-solvents for a specific polymer.

Relative evaporation rate (RER) The rate at which a material will vaporize (evaporate, change from liquid to vapour) compared to the rate of vaporization of a specific known material.

Reproductive toxicity A hazard associated with some substances, which interfere in some way with normal reproduction. Such substances are called reprotoxic. They may adversely affect sexual function and fertility in adult males and females, as well as causing developmental toxicity in the offspring.

Safety Data Sheets (SDS) Safety Data Sheets (SDS) are important documents in the safe supply, handling and use of chemicals. They help ensure that those who use chemicals in the workplace use them safely without risk of harm to users or the environment. They also provide a specification to help produce risk assessments (COSHH). They are not risk assessments.

Solubility The property of a solid, liquid or gaseous chemical substance (called solute) to dissolve in a solid, liquid or gaseous solvent. Solubility is defined as the maximum amount of a substance that will dissolve in a given amount of solvent at a specified temperature.

Solvent Star A method for assessing and documenting solvent tests for safe, efficient solvent use and the application of sustainable solvents in conservation (see Fife 2020).

Teas chart A graphical representation of fractional parameters derived from Hansen solubility parameters.

TMO (2,2,5,5-tetramethyloxolane) A non-peroxide forming ether solvent that can replace common hydrocarbon solvents such as toluene.

Toxicology Toxicology is the study of harmful effects of substances on people, animals and other living organisms.

Vapour pressure Vapour pressure or equilibrium vapour pressure is defined as the pressure exerted by a vapour in thermodynamic equilibrium with its condensed phases (solid or liquid) at a given temperature in a closed system. Related to the tendency of particles to escape from the liquid (or a solid) the equilibrium vapour pressure is an indication of a liquid's evaporation rate.

Volatile organic compound (VOC) VOCs are organic compounds with a vapour pressure of 0.01 KPa or more at the temperature of 293.15 K. They regroup multiple substances that are not only biogenic (natural) but also anthropogenic. They can cause various environmental and health and safety problems (some VOCs are carcinogenic, mutagenic and/or reprotoxic). They can have a direct impact on ozone creation in the atmosphere which is also responsible for global warming.

Bibliography

Abbott, S. 2013. *Hansen Solubility Parameters.*Available at: https://www.stevenabbott.co.uk/_downloads/Solubility%20Science%20Principles%20and%20Practice.pdf (accessed 20 August 2021).

Abbott, S. 2017. *Solubility Science: Principles and Practice.* Available at: https://www.stevenabbott.co.uk/_downloads/Solubility%20Science%20Principles%20and%20Practice.pdf (accessed 20 August 2021).

Abbott, S. 2021 *Practical Solubility – HSP.* Available at: https://www.stevenabbott.co.uk/practical-solubility/hsp-basics.php (accessed 5 October 2021)

Abou-Shehada, S., Clark, J.H., Paggiola, G. and Sherwood, J. 2016. 'Tunable solvents: shades of green', *Chemical Engineering and Processing: Process Intensification* 99: 88–96.

Alder, C., Hayler, J., Henderson, R., Anikó, R., Shukla, L., Shuster, L. and Sneddon, H. 2016. 'Updating and further expanding Gsk's Solvent Sustainability Guide', *Green Chemistry* 18(13): 3879–90.

American Chemical Society 2018. Solvent Selection Tool ACS GCI-PR. Available at: https://www.acs.org/content/acs/en/greenchemistry/research-innovation/tools-for-green-chemistry/solvent-selection-tool.html (accessed 20 August 2021).

Anastas, P. and Warner, J. 1998. *Green Chemistry: Theory and Practice.* Oxford: Oxford University Press.

Andrews, L.S. and Snyder, R. 1986. 'Toxic effects of solvents and vapors', in J. Doull (ed.), *Cassarett and Doull's Toxicology: The Basic Science of Poisons,* 3rd edn. New York: Macmillan.

Anon. 1854. *Mackenzie's Five Thousand Receipts in all the Useful and Domestic Arts.* Philadelphia: Hayes & Zell.

Ashcroft, C.P., Dunn, P.J., Hayler, J.D. and Wells, A.S. 2015. 'Survey of solvent usage in papers published', *Organic Process Research & Development 1997–2012* 19: 740–47.

Babu, N.S. and Reddy, S.M. 2014. 'Impact of solvents leading to environmental pollution', *Journal of Chemical and Pharmaceutical Sciences*. Available at: https://www.jchps.com/specialissues/Special%20issue3/06%20jchps%20si3%20nanni%2049-52.pdf (accessed 18 July 2021).

Baij, L., Hermans, J., Ormsby, B., Noble, P., Iedema, P. and Keune, K. 2020. 'A review of solvent action on oil paint', *Heritage Science* 8(1): art. no. 43.

Banti, D., Bonaduce, I., La Nasa, J., Lee, J., Lluveras-Tenorio, A., Modugno, F., Ormsby, B. and van den Berg, K. 2019. 'The influence of metal stearates on the water', in K.J. van den Berg, I. Bonaduce, A. Burnstock, B. Ormsby, M. Scharff, L. Carlyle, G. Heydenreich and K. Keune (eds), *Conservation of Modern Oil Paintings*. Cham: Springer.

Barker, R. and Ormsby, B. 2015. 'Conserving Mark Rothko's *Black on Maroon 1958*: the construction of a representative sample and the removal of graffiti ink', *Tate Papers* No. 23. Available at: http://www.tate.org.uk/research/publications/tate-papers/conserving-mark-rothkos-black-on-maroon-1958-construction (accessed 21 May 2021).

Bégin, D., Couture, C., Gérin, M. and Debia, M. 2020. *Solvants Verts: Fondements, Santé, Sécurité, Environnement Et Substitution*. Rapports Scientifiques, R-1089. Montréal, QC, CA: Institut de recherche Robert-Sauvé en santé et en sécurité du travail.

Breeden, S.W., Clark, J.H., Macquarrie, D.J. and Sherwood, J. 2012. 'Green solvents', in W. Zhang and B.W. Cue Jr. (eds), *Green Techniques for Organic Synthesis and Medicinal Chemistry*. Chichester: Wiley, 241–61.

Buhler, D.R. and Reed, D.J. 1990. *Ethel Browning's Toxicity and Metabolism of Industrial Solvents. Volume 2, Nitrogen and Phosphorus Solvents*. Amsterdam: Elsevier Science Publishers.

Burke, J. 1984. 'Solubility parameters: theory and application', *The Book and Paper Group Annual, The American Institute for Conservation* 3. Available at: https://cool.culturalheritage.org/byauth/burke/solpar/solpar7.html. (accessed 20 August 2021).

Buys, S. and Oakley, V. 1993. *The Conservation and Restoration of Ceramics*. Oxford: Butterworth-Heinemann.

Byrne, F., Saimeng, J., Paggiola, G., Petchey, T., Clark, T., Farmer, T., Hunt, A., McElroy, C., Sherwood, J. 2016. 'Tools and techniques for solvent selection: green solvent selection guides', *Sustainable Chemical Processes* 4(1): 1–24.

Byrne, F., Farmer, T., Clark, J., Hunt, A., Forier, B., Bossaert, G. and Hoebers, C. 2017. '2,2,5,5-tetramethyltetrahydrofuran (Tmthf): a non-polar, non-peroxide forming ether replacement for hazardous hydrocarbon solvents', *Green Chemistry* 19(15): 3671–8.

Cannon, A. 2011. 'Major developments in adhesive manufacture, 1750–2000', paper presented at the *Adhesives and Consolidants for Conservation: Research and Applications (CCI Symposium 2011).*

Capello, C., Ulrich, F. and Hungerbühler, K. 2007. 'What is a green solvent? A comprehensive framework for the environmental assessment of solvents', *Green Chemistry* 9(9): 927–34.

Caple, C. 2000. *Conservation Skills: Judgement, Method and Decision Making*. London: Routledge.

Clark, J.H. and Tavener, S.J. 2007. 'Alternative solvents: shades of green', *Organic Process Research & Development* 11(1): 149–55. Available at: https://doi.org/10.1021/op060160g.

Clark, J.H., Farmer, T.J., Hunt, A.J. and Sherwood, J. 2015. 'Opportunities for bio-based solvents created as petrochemical and fuel products transition towards renewable resources', *International Journal of Molecular Sciences* 16: 17101–59.

Conservation Unit Museums and Galleries Commission 1992. *Science for Conservators*, Volume 3. London: Routledge in association with the Conservation Unit Museums and Galleries Commission.

Constable, D.J.C., Jimenez-Gonzalez, C. and Henderson, R.K. 2007. 'Perspective on solvent use in the pharmaceutical industry', *Organic Process Research & Development* 11: 133–7.

Conti, A. 2007. *A History of the Restoration and Conservation of Works of Art*. Oxford: Butterworth-Heinemann.

De Piles, M. 1776. 'Élemens de peinture pratique'. *Oeuvres diverses de M. de Piles*, Volume 3. Amsterdam and Leipzig: Arkstée et Merkus.

de Silva, M. and Henderson, J. 2011. 'Sustainability in conservation practice', *Journal of the Institute of Conservation* 34(1): 5–15.

Doble, M. and Kruthiventi, A.K. 2007 'Alternate solvents', in *Green Chemistry and Engineering*. London: Academic Press, 93–104. Available at: https://

www.sciencedirect.com/science/article/pii/B9780123725325500067 (accessed 10 August 2021).

Dossie, R. 1764. *The Handmaid to the Arts*. London: J. Nourse.

Duquenoy-Bizouerne, A. and Falcy, M. 2003. 'Restaurateurs de tableaux: évaluation des risques toxicologiques', *Documents pour le Médecin du Travail* 96: 419–40.

Earle, M.J. and Seddon, K.R. 2000. 'Ionic liquids green solvents for the future', *Pure and Applied Chemistry* 72: 1391–8.

Enviro Tech 2018. Basic Industrial Cleaning – Types of Chemical Solvents. Available at: https://www.envirotechint.com/blog/basic-industrial-cleaning-types-of-chemical-solvents/ (accessed 24 October 2021).

EUR-Lex 1999. Council Directive 1999/13/EC of 11 March 1999 on the limitation of emissions of volatile organic compounds due to the use of organic solvents in certain activities and installations. Available at: https://eur-lex.europa.eu/legal-content/EN/ALL/?uri=celex%3A31999L0013 (accessed 10 August 2021).

EUR-Lex 2006. Regulation (EC) No 1907/2006 of the European Parliament and of the Council of 18 December 2006 Concerning the Registration, Evaluation, Authorization and Restriction of Chemicals (REACH), Establishing a European Chemicals Agency, Amending Directive 1999/45/EC and Repealing Council Regulation (EEC) No 793/93 and Commission Regulation (EC) No 1488/94 as well as Council Directive 76/769/EEC and Commission Directives 91/155/EEC, 93/67/EEC, 93/105/EC and 2000/21/EC. Available at: http://eur-lex.europa.eu/legal-content/EN/TXT/?uri=CELEX:02006R1907-20140410 (accessed 10 August 2021).

EUR-Lex 2008. Regulation (EC) No 1272/2008 of the European Parliament and of the Council of 16 December 2008 on Classification, Labelling and Packaging of Substances and Mixtures, Amending and Repealing Directives 67/548/EEC and 1999/45/EC, and Amending Regulation (EC) No 1907/2006. Available at: http://eur-lex.europa.eu/legal-content/en/ALL/?uri=CELEX:32008R1272 (accessed 10 August 2021).

EUR-Lex 2010. Directive 2010/75/EU of the European Parliament and of the Council of 24 November 2010 on Industrial Emissions (Integrated Pollution Prevention and Control). Available at: https://eur-lex.europa.

eu/legal-content/EN/ALL/?uri=CELEX%3A32010L0075 (accessed 10 august 2021).

European Chemicals Agency 2015a. Information on Chemicals. Available at: https://echa.europa.eu/information-on-chemicals (accessed 26 October 2021).

European Chemicals Agency 2015b. Guidance on REACH. Available at: http://echa.europa.eu/guidance-documents/guidance-on-reach (accessed 10 August 2021).

European Chemicals Agency 2015c. List of Restrictions. Available at: http://echa.europa.eu/addressing-chemicals-of-concern/restrictions/list-of-restrictions (accessed 10 August 2021).

European Chemicals Agency 2015d. Classification and Labelling Inventory. Available at: http://echa.europa.eu/information-on-chemicals/cl-inventory-database (accessed 11 August 2021).

European Chemicals Agency 2021. Available at: https://echa.europa.eu/ (accessed 8 November 2021).

European Commission 2015. Rapid Alert System for Dangerous Non-food Products (RAPEX). Available at: http://ec.europa.eu/consumers/consumers_safety/safety_products/rapex/index_en.htm (accessed 10 August 2021).

European Commission 2015. RAPEX Notification Reference A11/0116/14. Available at: http://ec.europa.eu/consumers/consumers_safety/safety_products/rapex/alerts (accessed 10 August 2021).

European Commission 2015. RAPEX Notification Reference A11/0079/15. Available at: http://ec.europa.eu/consumers/consumers_safety/safety_products/rapex/alerts (accessed 10 August 2021).

European Commission 2015. RAPEX Notification Reference A11/0072/15. Available at: http://ec.europa.eu/consumers/consumers_safety/safety_products/rapex/alerts (accessed 10 August 2021).

European Commission. 2015. RAPEX Notification reference A12/0840/15. Available at: http://ec.europa.eu/consumers/consumers_safety/safety_products/rapex/alerts (accessed 10 August 2021).

Fife, G.R. 2020. 'The Solvent Star: assessing and documenting solvent selection', *The Picture Restorer* 56: 41–4.

Finnish Safety and Chemicals Agency 2013. Toluene Substance Evaluation Report (under REACH). Available at: http://echa.europa.eu/documents/10162/a58633d6-1620-4764-b3bf-6308cad42e8b (accessed 10 August 2021).

Garcia, J.B. 2014. 'Cleaning areas: the location of tests in the cleaning of paintings', *International Journal of Conservation Science* 5(3): 283–94.

Gissi, A., Lombardo, A., Roncaglioni, A., Gadaleta, D., Mangiatordi, G.F., Nicolotti, O. and Benfenati, E. 2015. 'Evaluation and comparison of benchmark QSAR models to predict a relevant REACH endpoint: the bioconcentration factor (BCF)', *Environmental Research* 137: 398–409.

Hansen 2021. Hansen Solubility Parameters in Practice. Available at: https://www.hansen-solubility.com/HSPiP/ (accessed 20 August 2021).

Hedley, G. 1980. 'Solubility parameters and varnish removal a survey', *The Conservator* 4(1): 12–18. Available at: http://dx.doi.org/10.1080/01410096.1980.9994931.

Horie, C. 2010. *Materials for Conservation: Organic Consolidants, Adhesives and Coatings*, 2nd edn. Amsterdam: Butterworth-Heinemann.

Hossaini, R., Chipperfield, M.P., Montzka, S.A., Rap, A., Dhomse, S. and Feng, W. 2015. 'Efficiency of short-lived halogens at influencing climate through depletion of stratospheric ozone', *Nature Geoscience* 8: 186–90.

Hubbard, T. 1775. *Valuable Secrets Concerning Arts and Ttrades*. London: Will Haye

Hungerbühler, K. 2007. 'What is a green solvent? A comprehensive framework for the environmental assessment of solvents', *Green Chemistry* 9(9).

International Labour Organization 1971. Benzene Convention: Convention Concerning Protection against Hazards of Poisoning Arising from Benzene. Available at: http://www.ilo.org/dyn/normlex/en/f?p=NORMLEXPUB:12100:0::NO::P12100_ILO_CODE:C136 (accessed 10 August 2021).

ISO 14040:2006. Environmental Management – Life Cycle Assessment – Principles and Framework. International Organization for Standardization, Geneva, Switzerland. Available at: http://www.cscses.com/uploads/2016328/20160328110518251825.pdf (accessed 8 November 2021).

Jalan, I., Lundin, L. and van Stam, J. 2019. 'Using solubility parameters to model more environmentally friendly solvent blends for organic solar cell active layers', *Materials* 12: 3889.

Jessop, P. 2017. 'Green/alternative solvents', in M.A. Abraham (ed.), *Encyclopedia of Sustainable Technologies*. Amsterdam: Elsevier Science Publishers, 611–19.

Joshi, D.R. and Adhikari, N. 2019. 'An overview on common organic solvents and their toxicity', *Journal of Pharmaceutical Research International*: 1–18.

Keefe, M., Tucker, C., Behr, A.M., Meyers, G., Reinhardt, C., Boomgaard, T., Peitsch, C., Ormsby, B., Soldano, A., Phenix, A. and Learner, T. 2011. 'Art and industry: novel approaches to the evaluation and development of cleaning systems for artists' acrylic latex paints', *Coatings Tech* September: 30–43.

Klamt, A. 2011. 'The COSMO and COSMO-RS solvation models', *Wiley Interdisciplinary Reviews: Computational Molecular Science* 1(5): 699–709.

Liang, Q., Newman, P.A., Daniel, J.S., Reimann, S., Hall, B.D., Dutton, G. and Kuijpers, L.J.M. 2014. 'Constraining the carbon tetrachloride (CCl4) budget using its global trend and inter-hemispheric gradient', *Geophysical Research Letters* 41: 5307–15.

Lipshutz, B., Gallou, F. and Handa, S. 2016. 'Evolution of solvents in organic chemistry', *ACS Sustainable Chemistry & Engineering* 4(11): 5838–49.

Loubet, P., Tsang, M., Gemechu, E., Foulet, A. and Sonnemann, G. 2017. 'Life Cycle Assessment for green solvents', in F. Jérôme and R. Luque (eds), *Bio-Based Solvents*. Hoboken: Wiley. Available at: https://onlinelibrary.wiley.com/doi/10.1002/9781119065357.ch6 (accessed 10 August 2021).

Mackenzie, C. 1829. *Mackenzie's Five Thousand Receipts in All the Useful and Domestic Arts : Constituting a Complete Practical Library ... : A New American, from the Latest London Edition : With Numerous and Important Additions Generally : And the Medical Part Carefully Revised and Adapted to the Climate of the U. States : And Also a New and Most Copious Index*. Philadelphia: James Kay, Jun. and Brother.

Martinez, M.E., Demuth, P. and Ferreira, E.S.B. 2017. 'Preliminary solubility testing of green solvents for the removal and application of varnish on oil based paints'. Paper presented at the *Green Conservation of Cultural Heritage Palermo 2nd International Conference*, 16 November 2017.

Mills, J. and White, R. 1994. *The Organic Chemistry of Museum Objects*. Oxford: Butterworth-Heinemann.

Montabert, J.N.P. 1829. *Traité complet de la peinture*, Vol. 8. Paris: Bossange père.

Ormsby, B., Keefe, M., Phenix, A., von Aderkas, E., Learner, T., Tucker, C. and Kozak, C. 2016. 'Mineral spirits-based microemulsions: a novel cleaning system for painted surfaces', *Journal of the American Institute for Conservation* 55(1): 12–31.

Pagni, R.M. 2003. 'Ionic liquids as alternatives to traditional organic and inorganic solvents', in R.D. Rogers, K.R. Seddon and S. Volkov (eds), *Green Industrial Applications of Ionic Liquids*. NATO Science Series (Series II: Mathematics, Physics and Chemistry). Dordrecht: Springer, 92. Available at: https://doi.org/10.1007/978-94-010-0127-4_6.

Palmade-Ledantec, N. and Picot, A. 2010. 'La prévention des risques: le remplacement des solvants les plus toxiques par des solvants moins toxiques', *Conservation et restauration et sécurité des personnes*, 3–5 February 2010, Dragignan (France).

Patel, M. 1999. *Closing Carbon Cycles: Carbon Use for Materials in the Context of Resource Efficiency and Climate Change*. Wageningen: Ponsen and Looijen.

Pena-Pereira, F., Kloskowski, A. and Namieśnik, J. 2015. 'Perspectives on the replacement of harmful organic solvents in analytical methodologies: a framework toward the implementation of a novel generation of eco-friendly alternatives', *Green Chemistry* 17: 3687–705.

Phenix, A. 1997. 'Solvent abuse: some observations on the safe use of solvents in the cleaning of painted and decorated surfaces', *The Building Conservation Directory*. Available at: https://www.buildingconservation.com/articles/solvent/solvent.htm (accessed 20 August 2021).

Phenix, A. 2002. 'The swelling of artists' paints in organic solvents. Part 2, Comparative swelling powers of selected organic solvents and solvent mixtures', *Journal of the American Institute for Conservation* 41(1): 61–90.

Phenix, A. 2007. 'Generic hydrocarbon solvents: a guide to nomenclature', *WAAC Newsletter* 29(2).

Phenix, A. 2013. 'Effects of organic solvents on artists' oil paint films: swelling', in M.F. Mecklenburg, A.E. Charola and R.J. Koestler (eds), *New Insights into the Cleaning of Paintings*, Smithsonian Contributions to Museum Conservation No. 3. Washington, DC: Smithsonian Institution Scholarly Press, 69–76.

Phenix, A. and Sutherland, K. 2001. 'The cleaning of paintings: effects of organic solvents on oil paint films', *Studies in Conservation* 46(2): 47–60.

Pietsch, A. 2002. *Lösemittel: Ein Leitfaden für die restauratorische Praxis*. VDR Schriftenreihe zur Restaurierung Band 7, gebrauchtes Buch 2.

Pollet, P., Davey, E.A., Ureña-Benavides, E.E., Eckert, C.A. and Liotta, C.L. 2014. 'Solvents for sustainable chemical processes', *Green Chemistry* 16: 1034–55.

Prat, D., Pardigon, O., Flemming, H.-W., Letestu, S., Ducandas, V., Isnard, P., Guntrum, E., Senac T., Ruisseau, S., Cruciani, P. and Hosek, P. 2013. 'Sanofi's solvent selection guide: a step toward more sustainable processes', *Organic Process Research & Development* 17(12): 1517–25.

Prat, D., Hayler, J. and Wells, A. 2014. 'A survey of solvent selection guides', *Green Chemistry* 16: 4546–51.

Prat, D., Wells, A., Hayler, J., Sneddon, H., McElroy, C., Abou-Shehada, S. and Dunn, P. 2016. 'Chem21 selection guide of classical- and less-classical solvents', *Green Chemistry* 18(1): 288–96. Available at: https://pubs.rsc.org/en/content/articlepdf/2016/gc/c5gc01008j (accessed 26 October 2021).

Prati, S., Sciutto, G., Volpi, F., Vurro, R., Mazzeo, R., Rehorn, C., Blumich, B. *et al.* 2019. 'Cleaning oil paintings: NMR Relaxometry and SPME to evaluate the effects of green solvents and innovative green gels', *New Journal of Chemistry* 43(21): 8229–38.

Rivers, S. and Umney, N. 2003. *Conservation of Furniture*, 1st edn. Oxford: Butterworth Heinemann.

Sherman, J., Chin, B., Huibers, P.D.T., Garcia-Valls, R. and Hatton, T.A. 1998. 'Solvent replacement for green processing', *Environmental Health Perspectives* 106 (Suppl 1): 253–71.

Sherwood, J., De bruyn, M., Constantinou, A., Moity, L., McElroy, C.R., Farmer, T.J., Duncan, T., Raverty, W., Hunta, A.J. and Clark, J.H. 2014. 'Dihydrolevoglucosenone (Cyrene) as a bio-based alternative for dipolar aprotic solvents', *Chemical Communications* 50: 9650–52.

Sicaire, A.G., Vian, M., Fine, F., Joffre, F., Carré, P., Tostain, S. and Chemat, F. 2015. 'Alternative bio-based solvents for extraction of fat and oils: solubility prediction, global yield, extraction kinetics, chemical composition and cost of manufacturing', *International Journal of Molecular Science* 16: 8430–53.

Sigma-Aldrich 2021. Greener Products & Solutions. Available at: https://www.sigmaaldrich.com/GB/en/life-science/ssbi/greener-products-solutions (accessed 24 October 2021).

Smith, M., Jones, N., Page, S. and Peck Dirda, M. 1984. 'Pressure-sensitive tape and techniques for its removal from paper', *Journal of the American Institute for Conservation* 23(2): 101–13.

Soh, S. and Eckelman, M.J. 2016. 'Green solvents in biomass processing', *ACS Sustainable Chemistry and Engineering* 4(11): 5821–37.

Spurgeon, A. 2006. 'Watching paint dry: organic solvent syndrome in late-twentieth-century Britain', *Medical History* 50(2): 167–88.

Stavroudis, C. and Blank, S. 1989. 'Solvents and sensibility', *Western Area Art Conservation Newsletter* 11(2): 2–10.

StiCH 2021. Tools for Educated Sustainable Choices. Available at: https://stich.culturalheritage.org/ (accessed 24 October 2021).

Stavroudis, C. and Doherty, T. 2013. 'The Modular Cleaning Program in practice: application to acrylic paintings', in M.F. Mecklenburg, A.E. Charola and R.J. Koestler (eds), *New Insights into the Cleaning of Paintings*, Smithsonian Contributions to Museum Conservation No. 3. Washington, DC: Smithsonian Institution Scholarly Press, 139–45.

Stols-Witlox, M. 2011. 'Historical recipes for the cleaning of paintings 1600–1900', in *Preprints of the ICOM-CC 16th Triennial Meeting Lisbon, Portugal*. Paris: ICOM, 19–23.

Stols-Witlox, M. 2019. 'Chemists and historical recipes to clean paintings', *ARTECHNE*. Available at: https://artechne.wp.hum.uu.nl/chemists-and-historical-recipes-to-clean-paintings/ (accessed 20 August 2021).

Stoner, J. and Rushfield, R. 2012. *The Conservation of Easel Paintings*. London: Routledge.

Sutherland, K. 2013. 'Solvent leaching effects on aged oil paints', in M.F. Mecklenburg, A.E. Charola and R.J. Koestler (eds), *New Insights into the Cleaning of Paintings*, Smithsonian Contributions to Museum Conservation No. 3. Washington, DC: Smithsonian Institution Scholarly Press, 45–9.

Tebby, C., Mombelli, E., Pandard, P. and Péry, A.R.R. 2011. 'Exploring an ecotoxicity database with the OECD (Q)SAR Toolbox and DRAGON descriptors in order to prioritise testing on algae, daphnids, and fish', *The Science of the Total Environment* 409: 3334–43.

Timár-Balázsy, A. and Eastop, D. 1998. *Chemical Principles of Textile Conservation*. Oxford: Butterworth-Heinemann.

Tomei, F., Baccolo, T., Papaleo, B., Biagi, M., Signorini, S., Persechino, B. and Rosati, M. 1996. 'Effects of low-dose solvents on the blood of art restorers', *Journal of Occupational Health* 38(4): 190–95.

Torraca, G. 1980. *Solubilite et solvants utilises pour la conservation des biens culturels*. Rome: ICCROM.

UK Government 1990. Environmental Protection Act 1990. Available at: http://www.legislation.gov.uk/ukpga/1990/43/contents (accessed 14 August 2021).

UK Government 2005. The Hazardous Waste (England and Wales) Regulations 2005. Available at: http://www.legislation.gov.uk/uksi/2005/894/contents (accessed 23 October 2021).

United Nations 2017. Resolution adopted by the General Assembly on 6th July 2017. Work of the Statistical Commission pertaining to the 2030 Agenda for Sustainable Development (A/RES/71/313). Available at: https://web.archive.org/web/20201128194012/https:/undocs.org/A/RES/71/313 (accessed 10 August 2021).

United Nations Environment Programme 1987. The Montreal Protocol on Substances that Deplete the Ozone Layer. Available at: http://ozone.unep.org/en/treaties-and-decisions/montreal-protocol-substances-deplete-ozone-layer (accessed 10 August 2021).

United States Environmental Protection Agency 2006. *Life Cycle Assessment: Principles and Practice*. Cincinnati, Ohio: 3–9.

United States Environmental Protection Agency 2021. Green Chemistry. Available at: http://www.epa.gov/greenchemistry (accessed 24 October 2021).

Vibert, J. 1891. *La science de la peinture*. Paris: Paul Ollendorff.

Vila, A.S. and García, J.M.B. 2013. 'Computer applications and cleaning: Teas Fractional Solubility Parameter System in conservation', in M.F. Mecklenburg, A.E. Charola and R.J. Koestler (eds), *New Insights into the Cleaning of Paintings*, Smithsonian Contributions to Museum Conservation No. 3. Washington, DC: Smithsonian Institution Scholarly Press, 35–8.

Virot, M., Tomao, V., Ginies, C. and Chemat, F. 2008. 'Total lipid extraction of food using d-limonene as an alternative to n-hexane', *Chromatographia* 68: 311–13.

Welton, T. 2015. 'Solvents and sustainable chemistry', *Proceedings of the Royal Society A: Mathematical, Physical and Engineering Sciences* 471(2183). Available at: https://royalsocietypublishing.org/doi/full/10.1098/rspa.2015.0502 (accessed 24 October 2021).

Wills, S.T., Ormsby, B.A., Keefe, M.H. and Sammler R.L. 2021. 'Key characterization efforts to support the conservation and long-term care of Mark Rothko's painting *Black on Maroon 1958*', submitted to *Analytical Chemistry*, 6 November 2021.

Wolbers, R. 2000. *Cleaning Painted Surfaces: Aqueous Methods*. London: Archetype Publications.

World Health Organization 2015. IARC Monographs on the Evaluation of Carcinogenic Risks to Humans. Available at: https://monographs.iarc.who.int/ (accessed 23 October 2021).

Zumbühl, S. 2019. *Solvents, Solvation, Solubilization and Solution: The Solubility of Materials. An Introduction for Conservators including Solubility Data of Selected Conservation Materials*. Bern: HDW Publications. Available at: https://www.researchgate.net/publication/333405291 (accessed June 2021).

Zumbühl, S., Ferreira, E.S.B., Scherrer, N.C. and Schaible, V. 2012. 'The nonideal action of binary solvent mixtures on oil and alkyd paint: influence of selective solvation and cavitation energy', in M.F. Mecklenburg, A.E. Charola and R.J. Koestler (eds), *New Insights into the Cleaning of Paintings*, Smithsonian Contributions to Museum Conservation No. 3. Washington, DC: Smithsonian Institution Scholarly Press, 97–105.

www.ingramcontent.com/pod-product-compliance
Ingram Content Group UK Ltd.
Pitfield, Milton Keynes, MK11 3LW, UK
UKHW050142130426
469754UK00002B/28

9 781909 492844